MW00965192

Economic Incentives *for*
STORMWATER CONTROL

Edited by
Hale W. Thurston

CRC Press
Taylor & Francis Group
Boca Raton London New York

CRC Press is an imprint of the
Taylor & Francis Group, an **informa** business

Contents

Preface

Dealing with stormwater runoff in urban areas is a huge problem that is getting bigger and more expensive. As we cover porous surfaces with impervious structures—commercial buildings, parking lots, roads, and houses—finding places for rainwater and snowmelt to soak in becomes harder. This increases "flashy" runoff, causes lower groundwater tables, scrubs pollutants into streams, and creates flooding problems. The high speed "first flush" of water from concrete into local waterways destroys natural stream configurations by straightening them out, and washes away in-stream habitat for any number of organisms. In many Midwestern municipalities, the large quantity of stormwater runoff combining with sewage in underground pipes leads to combined sewer overflows, a particularly nasty situation where raw sewage is dumped into local rivers and streams. To deal with the problems caused by stormwater runoff, cities nationwide have adopted stormwater "best management practices," an interesting term, often erroneous, but established in the parlance nevertheless. Best management practices (BMP) come in all shapes and sizes, ranging from the so-called nonstructural practices such as educating homeowners to water the grass in the evening when it evaporates less, to aerated detention ponds you see in and around most new housing developments these days.

Getting commercial properties to participate in stormwater runoff control is usually done through command-and-control policy. That is, a regulating body at the federal, state, or local level has the authority to legally require commercial owners to install detention ponds or other runoff reducing practices. In many newer residential properties the case is similar, local governance requires a level of stormwater control to be built into the subdivision. It is often incumbent upon the developer, for example, to devote a certain area of the subdivision to a detention pond or other stormwater runoff capturing device. In older neighborhoods, however, property rights issues and lack of impervious surface restrictions conspire to cause built residential property to be one of the hardest-to-control sources of stormwater runoff.

Traditional stormwater control policies have concentrated on solutions that build centralized capacity to direct and hold excess runoff within a storm sewer system. However, such centralized infrastructures do not sufficiently alleviate water quality problems for small streams, and can be prohibitively expensive for municipalities that must work within budget constraints. It has been proposed by many landscapers, architects, planners, and others that perhaps the use of "green" localized management practices, such as rain gardens, could work as well as traditional "gray" pipes and basins at reducing the effects of stormwater runoff, and do so in a way that is more attractive in the landscape and possibly less expensive. This is the research question that many of us have asked ourselves. Is there a way to test this idea? Can we combine decentralized stormwater retention incented by a market mechanism, and study the hydrological benefits of BMPs distributed at the watershed scale, as well as evaluate the costs of meeting environmental quality standards?

Researchers have found that homeowners respond favorably to campaigns that offer economic incentives to participate in stormwater reducing behavior. They have also found that people respond to education about stormwater issues instead of monetary remuneration. Under most circumstances campaigns that rely on educational material alone will be less costly than policies that require monetary incentives.

Other research of economic interest looks into the extent to which people participate in such an environmental program for their own private gain and the extent to which they participate in order to contribute to the public good. It is argued that green infrastructure can provide benefits to those who possess it in the form of higher property values and aesthetics. It is also argued that green infrastructure provides benefits to those who do not possess it: a passerby, for example, can enjoy the birds and butterflies in a rain garden that is not his own. People who live downstream from an area that has received large amounts of stream-improving green infrastructure will reap benefits in the form of less flooding and higher quality water.

Some of the chapters in this book look explicitly at green infrastructure issues and some are more focused on the costs, benefits, and legal underpinnings of traditional stormwater control methods. Taken together these chapters raise and answer—or suggest how to answer—many of the questions facing those concerned with stormwater runoff issues.

I would like to thank Bill Lucas, Steve Buchberger, Bill Shuster, and Herb Cabezas for scientific guidance. I'd also like to thank Irma Shagla-Britton for her patience, and Kari Budyk and Arlene Kopeloff for oversight.

About the Editor

Hale W. Thurston is an economist in the U.S. Environmental Protection Agency's National Risk Management Research Laboratory in Cincinnati, Ohio. He received his PhD in economics from the University of New Mexico, a Master's in international affairs from Ohio University, and a Bachelor's degree in English literature from Bates College. His research currently focuses on nonmarket valuation of natural resources in the policy arena and the use of market incentives to promote low-impact development. He has been especially active in a research study that looks at the use of rain gardens and rain barrels to reduce the impacts of stormwater runoff. Dr. Thurston worked on a reforestation campaign and a beekeeping project in the Peace Corps in the Dominican Republic in the late 1980s. He currently resides in Cincinnati with his wife and two fantastic kids.

Contributors

Amy W. Ando
Department of Agricultural &
    Consumer Economics
University of Illinois
Urbana, Illinois

Christopher T. Behr
HDR Decision Economics
Silver Spring, Maryland

John B. Braden
Department of Agricultural &
    Consumer Economics
University of Illinois
Urbana, Illinois

Janet Clements
Stratus Consulting, Inc.
Washington, DC

W. Bowman Cutter
Department of Economics
Pomona College
Claremont, California

Andrew J. Erickson
Department of Civil Engineering
University of Minnesota
Minneapolis, Minnesota

Ahjond S. Garmestani
Office of Research and
    Development
National Risk Management
    Research Laboratory
U.S. Environmental Protection
    Agency
Cincinnati, Ohio

Haynes C. Goddard
Sustainable Technology Division
National Risk Management
    Research Laboratory
Office of Research and
    Development
U.S. Environmental Protection
    Agency
Cincinnati, Ohio

John S. Gulliver
Department of Civil Engineering
University of Minnesota
Minneapolis, Minnesota

Lisa Hair
Office of Watersheds, Oceans, and
    Wetlands
U.S. Environmental Protection
    Agency
Washington, DC

Theresa Hoagland
Formerly affiliated with
Sustainable Technology Division
Office of Research and
    Development
U.S. Environmental Protection
    Agency
Cincinnati, Ohio

Chelsea Hodge
Department of Economics
Pomona College
Claremont, California

J. Walter Milon
Department of Economics
University of Central Florida
Orlando, Florida

Franco A. Montalto
eDesign Dynamics LLC
New York, New York
and
Department of Civil, Architectural,
    and Environmental Engineering
Drexel University
Philadelphia, Pennsylvania

Punam Parikh
Formerly affiliated with
Sustainable Technology Division
Office of Research and
    Development
U.S. Environmental Protection
    Agency
Cincinnati, Ohio

Joanna Pratt
Stratus Consulting, Inc.
Washington, DC

David Scrogin
Department of Economics
University of Central Florida
Orlando, Florida

William Shuster
ORD-NRMRL
U.S. Environmental Protection
    Agency
Cincinnati, Ohio

Michael A. Taylor
Department of Political Science
    and Public Affairs
Seton Hall University
South Orange, New Jersey

Hale W. Thurston
ORD-NRMRL
U.S. Environmental Protection
    Agency
Cincinnati, Ohio

Patrick Walsh
National Center for Environmental
    Economics
U.S. Environmental Protection
    Agency
Washington, DC

Peter T. Weiss
Department of Civil Engineering
Valparaiso University
Valparaiso, Indiana

Ziwen Yu
Department of Civil, Architectural,
    and Environmental Engineering
Drexel University
Philadelphia, Pennsylvania

1

Background and Introduction

Hale W. Thurston

CONTENTS

The effects of stormwater runoff on stream ecosystems are exacerbated by urbanization and the coincident increase in impervious surface in watersheds. Proliferation of impervious surface creates higher peak flows that cause stream alteration and habitat degradation; it leads to reduced stream base flow, and in some areas results in toxic loading. Getting commercial properties to participate in stormwater runoff control is usually done through command-and-control tactics. That is, a regulating body at the federal, state, or local level has the authority to require commercial owners to install detention ponds or other runoff reducing practices. Residential properties pose a different problem: in new housing developments around the country, in part due to increased awareness of the inimical effects of stormwater runoff, municipalities often have enough public support to be able to require stormwater runoff best management practices (BMP). It is often incumbent upon the developer, for example, to devote a certain area of the subdivision to a detention pond or other stormwater runoff capturing device. In older neighborhoods, however, property rights issues and lack of impervious surface restrictions conspire to cause built residential property to be one of the hardest to control sources of stormwater runoff (Roy et al., 2006, and Parikh and her coauthors in Chapter 8 in this book). These properties cause many stormwater problems. Communities around the country are eager to come up with workable stormwater runoff solutions that are cost effective.

This book is concerned with the true costs and benefits of different stormwater practices being employed or envisioned around the country. By true costs and benefits, we mean among other things the market and nonmarket costs and benefits, the long-term operation and management costs and benefits, the opportunity costs of using land for stormwater runoff control, and the change in property values associated with stormwater

1

runoff practices and their subsequent impact on water quality. These true values need to be known in order to prescribe functioning programs that give an incentive for the adequate adoption of stormwater runoff management practices. These incentives are another focus of this book. For example, low-impact development or "green" infrastructure is being offered as an alterative to traditional or "gray" infrastructure as an aesthetically pleasing stormwater runoff control technique.

This book offers case studies of the application of various stormwater runoff control policies that have been used or suggested for use around the country, highlighting the economic aspects thereof. It also presents the theory behind the different mechanisms used, and illustrates successes and obstacles to implementation. Theoretically we would expect certain policies to be more effective at stormwater runoff control than others. We would expect command-and-control mechanisms to be the most effective at encouraging participation, that those programs with some requirement for participation would have higher effectiveness than those that were strictly voluntary, and for voluntary programs that offer economic incentives to be more effective than those programs not offering remuneration. Similarly we might expect certain technologies to perform better under certain circumstances. Taken together, the chapters in this book showcase the gaps between the theory and the application, and may help inform the political and legal realities on the ground to allow for the adoption of certain policies. The book also highlights the opportunities available to municipalities, stormwater managers, and stakeholder groups to enact sustainable, effective, stormwater management practices.

To see how this volume can be of assistance to stormwater decision makers, consider the following extended example: many large Midwestern cities are under a legal obligation with the U.S. Environmental Protection Agency (USEPA) to do something about their combined sewer overflow (CSO) problem. A CSO occurs when it rains heavily and too much water ends up in a system that was built for much smaller populations, thus sewage and rainwater mix, the sewage treatment plant cannot handle all the water, and, as a result, raw sewage is dumped into rivers. Local governments want to find inexpensive ways to comply with their obligations. One way is to build several more big best management practices, even including giant tunnels under the city like Chicago has to store the stormwater when it rains and let it into the sewer system slowly.

One problem with this solution is that it takes water out of natural streams. Another problem is that the expense is often politically prohibitive. The city of Cincinnati, for example, was considering a giant tunnel solution and very quickly the price tag went from $600 million to over a billion dollars. It has been proposed in several places that installing enough dispersed "low tech," or "green" management practices such as rain gardens and rain barrels and other landscape features can do the job of the big infrastructure projects at a fraction of the cost. But how effective are these practices? How much do they really cost? What are the benefits? How does the sewer district incent private property owners to install these low-tech devices in their lawns and around their property? We assume there needs to be offered some type of incentive for enough people to do it to make it work, and these are the questions this book addresses. Other books focus on the engineering aspects of stormwater runoff control. This volume is primarily concerned with the sociodemographic and economic aspects of people's participation in stormwater runoff control. Experts have a good idea of what runoff control mechanisms work. This book focuses on how stormwater managers and municipal leaders can get people to employ these practices, and what their true costs and benefits are.

The book begins with four chapters that broadly look at costs and benefits of stormwater management practices. In Chapter 2, Weiss, Gulliver, and Erickson look at some of the more concrete costs and pollutant removal effectiveness associated with standard stormwater management practices. This first part of Chapter 2 focuses especially on long-term operation and management (O&M) costs and opportunity costs of land, very important aspects in the lifetime of a stormwater management practice. One of the first, and perhaps most fundamental, questions the authors look at is: what kind of stormwater management practice (SMP) or technique will be optimum for a specific project and location over the life of the project? The chapter is concerned with "the best use of current and future resources," hence its focus on long-term costs and net present value calculations. So, for example, will a bioretention pond, constructed wetland, or infiltration trench best utilize current and future resources? Or will a standard dry detention pond maximize net present value? The authors note at times a single geographic region will have a variety of different kinds of SMPs that are popularly used, depending on rainfall, soil characteristics, and the like. In order to select the optimum SMP, associated construction and O&M costs need to be estimated and refined as the project progresses. An accurate first estimate will reduce the magnitude of later changes to the project, and thereby improve the process of choosing the optimum SMP. The chapter also looks closely at predicted removal effectiveness for certain nutrients and pollutants, and is able to provide cost-per-unit removed by the specific practices.

In Chapter 3, Braden and Ando continue the examination of the costs and effectiveness of stormwater management practices in new construction, by looking into some of the less obvious costs and benefits of enhanced national requirements for on-site stormwater management. For several years, the USEPA has gone back and forth on whether to strengthen the regulation of stormwater flows from new development. Construction site stormwater flow can be managed by conventional means such as the straw bales we see at sites, or basins that detain and settle pollutants, with "active treatment systems" that increase settling rates. There is also the potential, say Braden and Ando, to use "low-impact development" (LID) style new construction that reduces hydrological disruptions and promotes infiltration. These approaches differ in short- and long-term performance for both water quality and hydrological stability. When measures that increase infiltration are included at the outset in designing and implementing development plans, they can reduce the land disturbances that give rise to construction-phase pollution while also protecting the predevelopment hydrology in ways that reduce long-term stormwater management costs. Braden and Ando note that use of LID can also reduce development costs, thereby putting downward pressure on prices and increasing housing availability and access. Their benefit and cost analysis not only uses novel techniques, but it suggests that LID approaches may produce benefits of the same magnitude as the costs. Braden and Ando use up-to-date methods of benefit estimation for water quality, and determine that household willingness to pay for improved water quality is relatively high.

Chapter 4 by Montalto, Behr, and Yu extends the investigation into LID, or green infrastructure (GI), focusing on some of the sources of uncertainty inherent in predicting the ability of GI to cost-effectively reduce urban runoff. The authors maintain that municipalities, water utilities, and other natural resource planners need to account for such uncertainty better before they can make GI a central component of their infrastructure programs and land use plans. Traditional infrastructure decision-making processes are not easily applied to this problem, in large part because of the decentralization implied by the GI approach. Montalto and his coauthors present a multistep approach that integrates stakeholder participation into the GI decision-making process. The chapter highlights the use of version 2.0 of the low-impact development rapid assessment (or LIDRA) model. The authors show how this Web-based planning model enables users to quantitatively consider the uncertainty associated with GI performance, cost, and adoption in evaluating the cost-effectiveness of GI as an urban runoff reduction measure. And they examine how the model can be incorporated into a structured five-step process for simulating the cost-effectiveness of community-supported GI scenarios.

The authors recognize that the unpredictability of human behavior adds to the uncertainty of GI effectiveness. They propose, in contrast to financial incentives such as those highlighted in other chapters, targeted and structured stakeholder involvement in the process to encourage private property owners' participation in GI projects.

In Chapter 5, Garmestani and his coauthors take an applied look at the costs and benefits of LID practices, and begin to talk about some of the programs that stormwater facilities and municipalities around the country have used to incentivize the adoption of LID. This chapter is especially valuable for its presentation of case studies, and the lessons learned and hurdles overcome by local stormwater managers who have promoted the use of LID. The chapter also looks at the idea of uncertainty surrounding the effectiveness of LID management practices, a theme that is carried on in future chapters.

The next two chapters look into the estimation of the costs and benefits of runoff control using a hedonic methodology. Hedonic estimation has a long history in the environmental economics literature and is particularly helpful in estimating the nonmarket benefits associated with impacts on housing and land prices due to some change in a local environmental asset. Walsh, Milon, and Scrogin note, at the outset of Chapter 6, that stormwater runoff in the state of Florida will soon face stricter regulatory controls. The authors also assume that costs associated with achieving the stricter controls will be substantial. As with any economic analysis though, it is necessary to recognize both costs and benefits. Chapter 6 then investigates one dimension of these benefits: appreciating property prices in urban housing markets. The authors estimate several different specifications of hedonic price models employed in water quality and proximity valuation studies, and test several hypotheses about the implicit value of water quality. The study indicates that homeowners do indeed understand the water quality impacts of better stormwater management; the chapter concludes that housing values increase significantly in the face of improved water quality due to better control of stormwater runoff.

In Chapter 7, Thurston estimates some of the benefits of stormwater management applications, and especially how these are affected by property values. This can become a very important aspect of the process. As is mentioned in the chapter, the opportunity cost of the land that goes into a stormwater detention practice can be the deciding factor in LID application. There is a land value above which it is not efficient to devote it to stormwater retention. Thurston uses hedonic estimation results he generates from housing data in Ohio to inform the effectiveness of incentive mechanisms that might be used to encourage homeowners to retain stormwater on their property. He looks at a "fee with rebate" approach and a tradable allowances approach. He models both by incorporating the value of the land (the opportunity cost of using the property as stormwater retention) as estimated, in the cost of the stormwater management practice. Thurston maintains that this gives a more accurate picture of the costs associated with runoff control.

The next three chapters are concerned more explicitly with the theory behind incentive mechanisms available to stormwater managers to encourage people to retain stormwater runoff. People are familiar with some economic incentive programs that they don't even recognize as such. The "bottle bill" that many states have is an example of a "fee and rebate" program. Everyone who buys a beverage in a can or bottle pays the fee, and those for whom it is worth it, collect the bottles and cans and get the rebate. There are also many examples more specific to stormwater around the country.

Parikh and her coauthors in Chapter 8 look at different incentives available to municipalities for stormwater runoff control and explicitly address the legal, economic, and hydrological issues associated with different incentive mechanisms. According to the writers, "development and implementation of market mechanisms and incentives to reduce stormwater runoff involves interdisciplinary considerations and issues." The chapter develops an interdisciplinary view of the stormwater runoff issue, beginning with a brief description of stormwater runoff management from a hydrological perspective. They then present a background on types of market instruments and their related incentives as possible approaches to reducing the risks associated with both the magnitude and frequency of recurrence for excess stormwater runoff flows. This is followed by an analysis of how the federal Clean Water Act and state water laws have dealt with stormwater issues. These perspectives and methods are synthesized to develop several stormwater management scenarios that include stormwater user fees, stormwater runoff charges, allowance markets, and voluntary offset programs. Each of these programs would likely incorporate stormwater best management practices at the watershed level, yet in different ways. The chapter concludes with a comparison of the various mechanisms.

In Chapter 9, Hodge and Cutter delve more deeply into a single market mechanism: the in-lieu fee (ILF). The idea of using the ILF option is popular among stormwater managers and municipal leaders because it can help mitigate the economic and political costs of on-site BMP requirements. The ILF allows developers and property owners to pay a fee in lieu of meeting the regulatory requirement. The revenue from the fees is used by the local government to provide runoff management elsewhere in the community. The authors note that ILFs have the potential to provide the control and predictability of a command-and-control policy but can take advantage of the flexibility and cost savings associated with incentive-based policies. Hodge and Cutter argue that when dealing with regional versus local stormwater control measures, it is key to recognize the difference between marginal and average costs. Small local management practices often appear expensive relative to a very large regional practice, but this is because the average costs are compared. When the marginal costs of increasing capacity are used, on-site management practices may compare favorably.

But Hodge and Cutter are also interested in the public acceptance of the ILF policy. They present the results of a nationwide survey of municipalities that had adopted ILF. They asked stormwater managers various questions about their ILF program and learned among other things that ILFs are not just adopted as a cost-effectiveness tool, but they can help convince the public to accept more stringent runoff requirements. The main reason for offering an ILF though, according to the survey, was the belief that in at least some circumstances regional stormwater management was more effective than on-site BMPs.

In Chapter 10 Goddard looks at the economic aspects of a cap-and-trade incentive mechanism. He focuses on the description and analysis of how a constructed pollution trading program, or cap-and-trade in popular parlance, might be applied to guiding a stormwater management program to achieve runoff reduction goals at least cost. The chapter provides a nice overview of the theory behind any cap-and-trade system and why they are looked upon favorably by economists. The intuition in the stormwater runoff setting, he notes, is relatively straightforward: one compares the cost of an individual gray or green investment to handle the next increment in runoff, and apply whichever is less costly. Goddard then digs into some of the specific issues applying cap-and-trade to the stormwater problem. Of particular interest to the reader is the treatment, within the cap-and-trade paradigm, of the stochastic nature of nonpoint-source stormwater runoff. The chapter also tackles the issues of transfer coefficients, covariances, and trading ratios, and accounts for these in the model. Goddard concludes the chapter with a brief overview of some of the more practical issues involved with the adoption of an allowance trading system, such as what kind of market structure (sole source offsets, clearinghouse) best facilitates a well-running market? And what are some of the steps necessary for a market to be established, such as the existence of a good GIS database, and a menu of approved management practices for an area?

Together these chapters offer an insight into some of the complexities of the costs, benefits, and incentives—in short, the economics—behind the use of different management practices for stormwater runoff control. Impervious surface area is not shrinking in our country, so the problems of stormwater runoff will only be increasing over time. Innovative solutions for those few places where there is actually a reduction in urban impervious surfaces, as in the re-purposing of urban land in places like Cleveland and Detroit, can draw from the same incentive mechanisms and economic analysis. It is up to stormwater managers, municipal leaders, and individual homeowners to know what the problem is and what some of the potential solutions are. Decisions concerning the types of stormwater management practices to use depend critically upon the costs of the various methods. The decisions on policy and chosen practices also depend upon the goals set forth by the community. A neighborhood may want the most cost-effective method or the program that maximizes net benefits. The goal may be to attain the most widespread public acceptance. No matter the goal, the chapters in this volume provide informed insight into the potential obstacles and solutions for effective stormwater management.

Disclaimer

Any opinions expressed in this book are those of the authors and do not, necessarily, reflect the official positions and policies of the EPA. Any mention of products or trade names does not constitute recommendation for use by the EPA.

Reference

Roy, Allison H., Heriberto Cabezas, Matthew P., Clagett, N., Hoagland, Theresa, Mayer, Audrey L., Morrison, Matthew A., Shuster, William D., Templeton, Joshua J., and Hale, W. Thurston. (2006). Retrofit stormwater management, *Stormwater*, 7(3): 16–29.

2

Costs and Effectiveness of Stormwater Management Practices

Peter T. Weiss, John S. Gulliver, and Andrew J. Erickson

CONTENTS

Introduction

As watershed districts, municipalities, stormwater engineers, and other stormwater professionals strive to manage stormwater in an optimum and cost-effective manner, they face myriad choices and design decisions. One of the first and perhaps most fundamental questions that must be answered is what kind of stormwater management practice (SMP) or technique will be optimum for a specific project and location over the life of the project? In this case, optimum means the best use of current and future resources. For example, will a bioretention pond, constructed wetland, or infiltration trench best use current and future resources? Alternatively, perhaps the installation of a standard dry detention pond will optimize resources. Many SMPs are available and often times a single geographic region will have a variety of different kinds of SMPs. To select the optimum SMP, associated construction and operating and maintenance (O&M) costs need to be estimated and refined as the project progresses. An accurate first estimate will reduce the magnitude of later changes to the project, and thereby improve the process of choosing the optimum SMP.

Overall construction costs will vary and will depend on land values, required site preparation (e.g., mulching and grubbing, tree removal, earthwork, etc.), and the specified design requirements of the particular SMP. O&M costs will depend on the type of SMP, the frequency and complexity of inspection and maintenance, watershed land uses, rainfall patterns and climate, and on the goals of the relevant stormwater management plan, all of which may vary more than variables related to construction costs. For example, based on a survey distributed to 108 municipalities throughout Minnesota, Kang et al. (2008) found that annual staff-hours spent on routine maintenance (e.g., mowing, visual inspection, trash removal, vegetation management, etc.) to be highest for rain gardens and wetlands and the lowest for dry ponds, filters, infiltration systems, and filter strips/swales (Figure 2.1). As staff-hours will affect O&M budgets over the life of the SMP, they must be considered when selecting a SMP.

The complexity of maintenance will affect required staff-hours and other items that will affect O&M expenses such as consultant fees, equipment rental, landfill costs (if applicable), and so on. Maintenance complexity will depend on the type of SMP, desired levels of SMP performance,

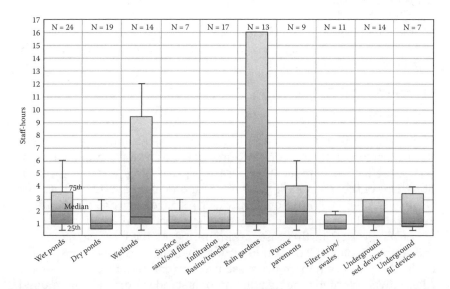

FIGURE 2.1

Staff-hours spent on routine maintenance actions in Minnesota. (From Kang, J.H., P.T. Weiss, J.S. Gulliver, and B.C. Wilson. Nov.–Dec. 2008. Maintenance of Stormwater BMPs: Frequency, Effort, Cost. *Stormwater*, 9(8), pp. 18–28.)

maintenance crew training, experience, and skills along with other variables.

Survey respondents (Kang et al., 2008) were asked to judge the complexity of the maintenance that was encountered in the field on a scale that ranged from *minimal* (a stormwater professional or consultant was seldom needed) to *simple* (stormwater professional or consultant is occasionally needed) to *moderate* (stormwater professional or consultant is needed about half the time for maintenance), and finally to *complicated* (stormwater professional or consultant is always needed). Survey results are shown in Table 2.1.

The variables discussed above, among others, will affect O&M costs for the lifetime of the SMP. If stormwater professionals are going to optimize the use of current and future resources, a thorough understanding of both construction and O&M requirements is essential. Furthermore, at the very least, accurate estimates of construction and O&M cost must be developed. This chapter presents a first step towards that goal.

Historical construction cost data (not including land costs) have been collected and analyzed to provide an estimate of SMP construction costs as a function of the stormwater runoff volume the SMP is designed to treat. Annual O&M costs have been estimated from expected maintenance schedules and have been combined with construction costs to estimate an overall total present SMP cost for a 20-year design life. As costs would be meaningless without a consideration of the performance of each SMP, the total amount of suspended sediment and phosphorus that is expected to be removed over the 20-year design life has been estimated for each kind of SMP.

Sufficient data were available for dry detention basins, wet retention basins, constructed wetlands, infiltration systems, and bioretention systems (which includes rain gardens), and sand filters (both above and underground). As the field grows, more and more SMPs are becoming available, including many commercially available proprietary devices. Although sufficient data were not available for these additional SMPs, they should be considered during the SMP selection process. Thus, as responsible stormwater professionals, the discipline must continue to expand and refine available estimates related to the cost and effectiveness of SMPs.

Cost Estimation

Based on published cost data of actual SMPs a method, which is described herein, was developed that will enable designers and planners to make

TABLE 2.1

Survey on Maintenance Complexity for Different Stormwater BMPs in Minnesota

	Maintenance Complexity[a]			
BMP Category	Minimal (%)	Simple (%)	Moderate (%)	Complicated (%)
Wet Ponds	55	32	5	9
Dry Ponds	65	30	0	5
Constructed Wetlands	40	13	40	7
Surface Sand/Soil Filters	63	0	25	13
Infiltration Basins/Trenches	33	40	13	13
Bioretention Practices	38	31	13	19
Porous Pavements	42	8	42	8
Filter Strips/Swales	62	31	0	8
Underground Sed. Devices	40	33	7	20
Underground Filt. Devices	50	20	10	20

Source: Kang et al. (2008) Maintenance of stormwater BMPs: Frequency, effort, cost. *Stormwater,* 9(8, Nov.–Dec.):18–28.

[a] *Minimal:* stormwater professional or consultant is seldom needed; *Simple:* stormwater professional or consultant is occasionally needed; *Moderate:* stormwater professional or consultant is needed about half the time for maintenance; *Complicated:* stormwater professional or consultant is always needed.

estimates of the total present cost (TPC) of various SMPs if the size of the SMP is known. The TPC is defined as the present worth of the total construction cost of the project plus the present worth of 20 years of annual O&M costs. The values reported do not include costs of pretreatment units (which may be required), design or engineering fees, permit fees, land costs, or contingencies, among others.

Water Quality Volume

The costs of SMP projects are often reported along with the corresponding watershed size (usually in acres or square feet) or the water quality volume (WQV) for which the SMP was designed. The water quality volume is usually defined as the volume (typically in acre feet or cubic feet) of runoff the SMP is designed to store and treat.

Claytor and Schueler (1996) calculate the WQV (ft^3) for a particular precipitation amount:

$$WQV = \left(\frac{43560}{12}\right) * P * R_V * A \qquad (2.1)$$

where P = precipitation depth (inches), R_V = ratio of runoff to rainfall in the watershed, A = watershed area (acres), and the constants are conversion factors.

The ratio of runoff to rainfall, R_V, has the most uncertainty of the parameters in Equation (2.1). For this analysis, a relatively simple relationship was used (Claytor and Schueler, 1996; Young et al., 1996),

$$R_V = 0.05 + 0.009 * (I) \qquad (2.2)$$

where I is the percentage (0–100) of the watershed that is impervious. Equation (2.2) indicates that, for a 100% impervious watershed, 95% of the rainfall becomes runoff.

Total Construction Costs

Values of total construction costs of SMPs throughout the United States were collected from published literature. Although data were collected on many SMP technologies, sufficient data to perform a cost analysis could be found only for dry detention basins, wet retention basins, constructed wetlands, infiltration trenches, bioinfiltration filters, sand filters, and swales. All data were adjusted to reflect costs in the upper Midwest of the United States (USEPA rainfall zone (1) by means of "Regional Cost Adjustment Factors" as reported by the United States Environmental Protection Agency (USEPA, 1999) and were adjusted to year 2005 dollars using an annual inflation rate of 3%. A value of 3% was chosen after an analysis of building cost indexes for an 11-year period from 1992 to 2003. (Turner Construction, 2004) revealed the average annual inflation was 3.26% with a range from 0.3 to 5.4%.

The cost data, which were collected, were analyzed as a function of the water quality volume for which the particular SMP was designed. When the cost data were converted to unit construction costs, defined as the total construction cost per acre of watershed or per cubic foot of WQV, the data, in all cases except for bioinfiltration filters, exhibited an "economy of scale." In other words, when the unit construction cost was plotted versus the size (i.e., watershed area or WQV), the unit cost tended to decrease as the size increased. As mentioned, the only exception to this trend was bioinfiltration filters, which exhibited a slight increase in unit cost with increasing size.

When comparing unit-cost data based on watershed area and WQV, the data based on WQV were, in most cases, observed to have less scatter as quantified by the standard error of the y-estimate. Thus, to provide for as much consistency as possible while minimizing scatter overall, WQV-based unit construction costs were selected for use over watershed area based unit construction costs. A swale is usually designed for a peak flow rate and could have a wide variety of lengths, thus basing a cost analysis on WQV is not practical. Instead, projected cost estimates per linear foot of swale as a function of geometry were collected and analyzed. Using these data, the cost per linear foot of a grassed/vegetative swale was found to be a function of the top width of the swale. Thus, a second method, used only to estimate the construction costs of swales, was developed and is based on construction cost per linear foot as a function of swale top width.

Figures 2.2 through 2.8 show the unit construction cost data analyzed in graphical form. Shown is the dashed, best-fit line through the data and the 67% confidence interval as shown by solid lines on either side of the best-fit line. The 67% confidence interval shows the bounds that will, on average, contain 67% of the data. If there are sufficient data (~20) and the distribution is, in this case, truly log-normal, then one-third of the data will fall outside the 67% confidence interval. The data originating from Brown and Schueler (1997) were read graphically whereas the values from SWRPC (1991), Caltrans (2004), and ASCE (2004) were

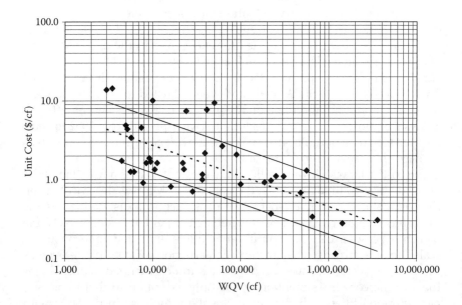

FIGURE 2.2
Unit construction costs of dry detention basins. (Data from Brown and Schueler, 1997. *The economics of stormwater BMPs in the mid-Atlantic region: Final report.* Center for Watershed Protection, Silver Spring, MD.; ASCE, 2004 International stormwater best management practices (SBMP) database. American Society of Civil Engineers; Caltrans, 2004. *BMP retrofit pilot program – Final report, Appendix C3.* California Department of Transportation, Division of Environmental Analysis, Sacramento, CA.)

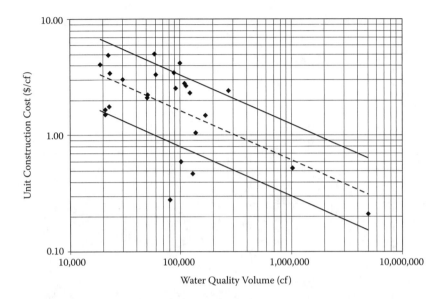

FIGURE 2.3
Unit construction costs of wet basins. (Data from Brown and Schueler, 1997. *The economics of stormwater BMPs in the mid-Atlantic region: Final report.* Center for Watershed Protection, Silver Spring, MD; Caltrans, 2004. *BMP retrofit pilot program – Final report, Appendix C3.* California Department of Transportation, Division of Environmental Analysis, Sacramento, CA.)

given in tabular form. The data from Caltrans (2004) were collected by means of a survey distributed by Caltrans to other agencies throughout the United States. It should be noted that the total construction costs of SMPs installed by Caltrans were available but these values were omitted from this analysis because their costs were typically about ten times higher than similarly sized projects constructed by other agencies. Caltrans attributed these high costs to the fact their projects were retrofits and were not original projects that were installed as part of larger construction projects.

The unit construction costs (with 67% confidence interval), excluding land costs, of each stormwater SMP except for grassed/vegetative swales can be described (fit) in equation form as

$$UCC = \beta_0 (WQV)^{\beta_1} \qquad (2.3)$$

where UCC = unit construction cost (2005 USEPA rainfall zone 1 dollar per cubic foot of WQV), WQV = water quality volume (ft^3), and β_0 and β_1 = constants.

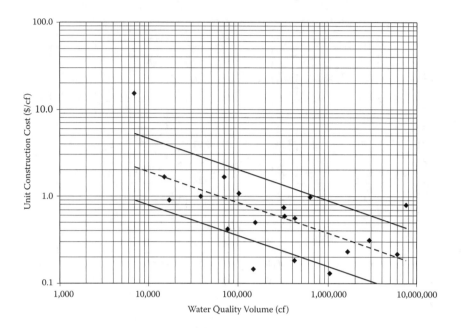

FIGURE 2.4
Unit construction costs of constructed wetlands. (Data from Brown and Schueler, 1997, *The economics of stormwater BMPs in the mid-Atlantic region: Final report.* Center for Watershed Protection, Silver Spring, MD; Caltrans, 2004. *BMP retrofit pilot program – Final report, Appendix C3.* California Department of Transportation, Division of Environmental Analysis, Sacramento, CA.; ASCE, 2004. International stormwater best management practices (SBMP) database. American Society of Civil Engineers.)

For each stormwater SMP the values of β_0 and β_1 for the average UCC, the values of β_0 and β_1 for the upper and lower 67% confidence intervals, and the range of WQV for which data were collected are given in Table 2.2.

Of the data collected for sand filters, some contained information on the type of sand filter (e.g., Austin or Delaware) whereas other data included no such description. Interestingly, when analyzing the sand-filter data for unit costs, there was essentially the same amount of scatter when the data of each sand-filter type were analyzed alone as there was when all sand-filter data were combined and analyzed together. This suggests that sand-filter unit-construction costs are independent of the type of filter and, as a result, cost estimates developed herein do not differentiate between sand-filter types. Figure 2.7 does differentiate between the Austin, Delaware, and undefined data by the data marker but, because no trend was observed for individual filter types, the best-fit line is drawn through the combined data.

The uncertainty observed in the construction cost data for all SMPs is most likely due to several factors such as design parameters, regulation

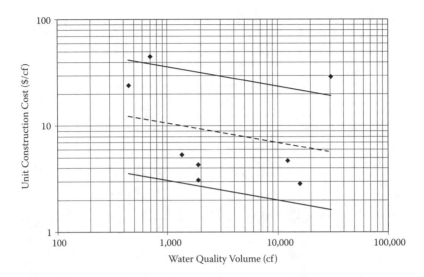

FIGURE 2.5
Unit construction costs of infiltration trenches. (Data from Caltrans, 2004. *BMP retrofit pilot program – Final report, Appendix C3.* California Department of Transportation, Division of Environmental Analysis, Sacramento, CA).

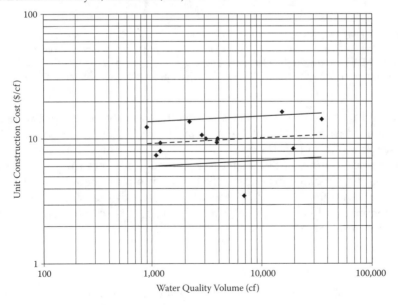

FIGURE 2.6
Unit construction costs of bioinfiltration filters. (Data from Brown and Schueler, 1997. *The economics of stormwater BMPs in the mid-Atlantic region: Final report.* Center for Watershed Protection, Silver Spring, MD; Caltrans, 2004. *BMP retrofit pilot program – Final report, Appendix C3.* California Department of Transportation, Division of Environmental Analysis, Sacramento, CA.)

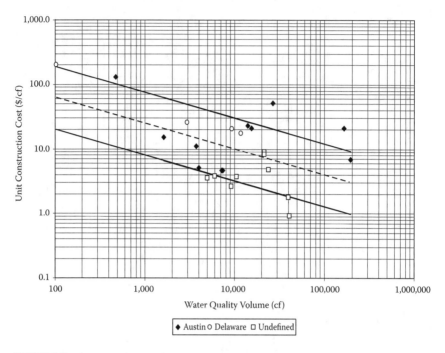

FIGURE 2.7
Unit construction costs of sand filters. (Data from Brown and Schueler, 1997; *The economics of stormwater BMPs in the mid-Atlantic region: Final report.* Center for Watershed Protection, Silver Spring, MD; Caltrans, 2004. *BMP retrofit pilot program – Final report, Appendix C3.* California Department of Transportation, Division of Environmental Analysis, Sacramento, CA.)

requirements, soil conditions, site specifics, and the like. For example, variable design parameters that would affect the total construction cost include pond side slopes, depth, and free board on ponds, total wet pond volume, outlet structures, the need for retaining walls, and so on. Site-specific variables include clearing and grubbing costs and fencing around the SMP among others. Due to the wide number of undocumented variables that affect the data, this scatter would be difficult to minimize.

Land-Area Requirements

An important cost of any SMP is that of the land area on which the SMP will be located. For urban areas, in which land is typically at a premium, this cost can be relatively large. On the contrary, in more open rural areas, land costs might be a very small percentage of the total project costs. Due to the extreme range of land costs and variability from site to site, no attempt was made to incorporate this cost into the cost analysis. The land area

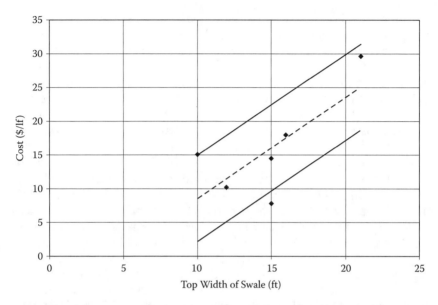

FIGURE 2.8
Unit construction costs of grassed/vegetative swales. (Data from SWRPC, 1991 Southeastern Wisconsin Regional Planning Commission. *Costs of urban nonpoint source water pollution control measures*. Waukesha: SWRPC.)

TABLE 2.2

Equation (2.3) Constants for UCC and Associated Confidence Intervals

Best Management Practice	Average Unit Construction Cost		Upper Confidence Interval		Lower Confidence Interval		Water Quality Volume Range
	β_0	β_1	β_0	β_1	β_0	β_1	(cubic ft)
Dry Basins	98.28	−0.389	219.39	−0.389	43.93	−0.389	3000–3,560,000
Wet Basins	215.28	−0.423	437.45	−0.423	105.94	−0.423	14,500–7,600,000
Sand Filters	392.64	−0.400	1,209.13	−0.400	127.51	−0.400	100–195,000
Bioretention Filters	6.70	0.045	10.04	0.045	4.47	0.045	900–35,00
Const. Wetlands	50.47	−0.355	121.05	−0.355	21.04	−0.355	7,000–7,600,000
Infiltration Trenches	37.50	−0.183	128.58	−0.183	10.94	−0.183	450–30,700

requirements, however, and therefore the associated land costs, of each SMP technology can vary dramatically and would, in many scenarios, have a significant impact on the total cost of a project. For example, a sand filter placed underground below a parking lot would, in effect, require no additional land area. A constructed wetland designed to treat the same volume of runoff as the underground sand filter, however, would require significant additional land area that may preclude the use of wetlands.

Given the variability of land costs and the variety of potential SMPs that could be used, the impact of land costs must be done on an individual, case-by-case basis. Table 2.3, which lists typical SMP land-area requirements for effective treatment, is presented to assist designers and planners in making such a comparison. Values reported in Table 2.3 by Claytor and Schueler (1996) are for the general category of SMP system and may include more than one specific type of SMP. For example, their pond category may include both wet and dry ponds. Table 2.4 lists wet pond areas required for control of solid particles that are 5 and 20 microns in size as reported by Pitt and Voorhees (1997). If the land costs in the locale of a particular project are known, these costs can be combined with the information presented in the tables to estimate a range of possible land area costs associated with each SMP under consideration. This information gives only a possible range of land area costs. For more accurate land area cost estimates, a preliminary SMP design is recommended.

TABLE 2.3

Reported SMP Land Area Requirements for Effective Treatment

SMP System	SMP Area (% of Impervious Watershed)[a]	SMP Area (% of Watershed)[b] Except as Noted
Bioretention	5	—
Wetland	3– 5	3– 5
Wet/Retention Basin	2–3	—
Sand Filter	0– 3	—
Dry Det Basin	—	0.5– 2.0[c]
Infiltration Trench	2– 3	—
Filter Strips	100	—
Swales	10–20	—
Pond	—	2– 3
Infiltration	—	2–3
Filter	—	2– 7

[a]*From USEPA. (1999). Preliminary data summary of urban stormwater best management practices. EPA-821-R-99-012, Washington, DC.*
[b]*From Claytor, R.A., and T.R. Schueler. (1996). Design of Stormwater Filtering Systems. Center for Watershed Protection, Silver Spring, MD.*
[c]*From Urban Drainage Flood Control District (UCFCD), Denver CO. (1992). Best Management Practices, Urban Storm Drainage Criteria Manual, Vol. 3, Denver.*

TABLE 2.4

Typical Land Area Requirements (% of Total
Watershed) for Wet Ponds (i.e., Wet Basins)

Land Use	5 Micron Control	20 Micron Control
100% Paved	3.0	1.1
Freeways	2.8	1.0
Industrial	2.0	0.8
Commercial	1.7	0.6
Institutional	1.7	0.6
Residential	0.8	0.3
Open Space	0.6	0.2
Construction	1.5	0.5

Source:. Pitt, R.E. and J.G. Voorhees. 1997. Storm water
quality management through the use of
detention basins, a short course on storm
water detention basin design basics by integrating water quality with drainage objectives. April 29–30 and May 21–22, University
of Minnesota, St. Paul, MN.

Note: As percentage of total watershed for wet ponds
(i.e.. wet basins).

Operating and Maintenance Costs

Over the lifetime of a SMP, the O&M costs can be a significant expense
that must be considered when selecting a treatment method. No data were
found, however, that documented actual O&M costs of existing SMPs. At
best, available data consisted only of expected or predicted O&M costs
of recently constructed SMP projects. Often times, general guidelines of
estimated annual O&M costs were presented as a percentage of the total
construction cost. For example, the USEPA (1999) gives a summary of typical SMP annual O&M costs as shown in Table 2.5. Included in the right
column of Table 2.5 is the range of the authors' collection of predicted
O&M costs as a percentage of the construction cost.

Ideally, the TPC estimate would be based on actual O&M costs of existing SMPs but, as mentioned, estimated annual O&M costs were the only
available data. When these data were evaluated to determine how the estimated O&M costs compared to those summarized by the USEPA, a trend
was observed for all SMPs except infiltration trenches in which the annual
O&M cost as a percentage of the construction cost decreased with increasing construction cost. The collected annual O&M cost data are shown as
log–log plots in Figures 2.9 through 2.15. As with the construction cost
data, the best-fit line through the data and the 67% confidence interval
are shown. Actual wet basin O&M cost data collected from the City of
Green Bay, Wisconsin, are shown in Figure 2.10. The Green Bay data have

TABLE 2.5

Typical Annual O&M Costs of SMPs

SMP	Summary of Typical AOM Costs (% of Construction Costs) (USEPA, 1999)	Collected Cost Data: Estimated Annual O&M Costs (% of Construction Costs)
Retention Basins and Constructed Wetlands	3–6	—
Detention Basins	<1	1.8–2.7
Constructed Wetlands	2	4–14.1
Infiltration Trench	5–20	5.1–126
Infiltration Basin	1–3 5–10	2.8–4.9
Sand Filters	11–13	0.9–9.5
Swales	5–7	4.0–178
Bioretention	5–7	0.7–10.9
Filter Strips	$320/Acre (maintained)	—
Wet Basins	Not Reported	1.9–10.2

a steeper slope than the estimated costs (i.e., the annual O&M cost as a percentage of the construction cost decreased at a more rapid rate) but does overlap with the range of the estimated costs.

The best fit and 67% confidence interval lines of Figures 2.9 through 2.15 were fit to Equation (2.4).

$$AOM = \beta_0 (TCC)^{\beta_1} \qquad (2.4)$$

where *AOM* = annual operating and maintenance costs as a percentage of total construction cost, *TCC* = total construction cost (2005 USEPA rainfall zone 1 dollars), and β_0 and β_1 = constants. In the following section the annual O&M costs are combined with the unit construction costs to develop an estimate for the total present cost of each SMP as a function of WQV or, in the case of swales, as a cost per linear foot as a function of swale top width. Table 2.6 lists the best-fit constants β_0 and β_1 for each of the SMPs shown in Figures 2.9 through 2.15.

Total Present Cost

If an estimate of the total construction cost of a SMP were desired, the data presented in Figures 2.2 through 2.7 could be used in a stand-alone manner simply by multiplying the unit construction cost ($/ft³) by WQV (ft³). The construction cost of swales could be estimated by multiplying the unit cost ($/ft) by the swale length (ft). A more practical estimate, however, is that of the total costs needed not only to construct but to maintain and

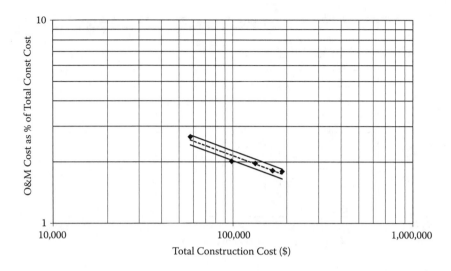

FIGURE 2.9
Annual O&M costs of dry detention basins. (Data from Landphair, H.C., J.A. McFalls, and D. Thompson, 2000. *Design methods, selections, and cost-effectiveness of stormwater quality structures.* Texas Transportation Institute, The Texas A&M University System, College Station, TX.)

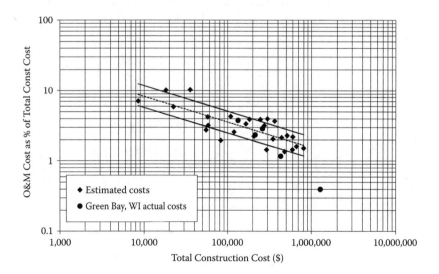

FIGURE 2.10
Annual O&M costs of wet basins. (Data from SWRPC, 1991 Southeastern Wisconsin Regional Planning Commission. *Costs of urban nonpoint source water pollution control measures.* Waukesha: SWRPC;. Wossink, A., and B. Hunt. 2003. *The economics of structural stormwater BMPs in North Carolina.* University of North Carolina Water Resources Research Institute report #UNC-WRRI-2003-344.)

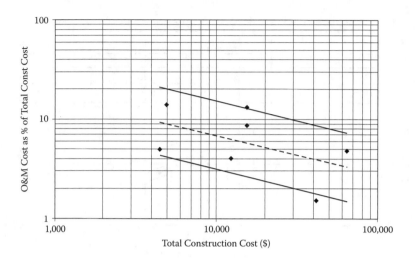

FIGURE 2.11
Annual O&M costs of constructed wetlands. (Data from Wossink, A., and B. Hunt. 2003. *The economics of structural stormwater BMPs in North Carolina.* University of North Carolina Water Resources Research Institute report #UNC-WRRI-2003-344.)

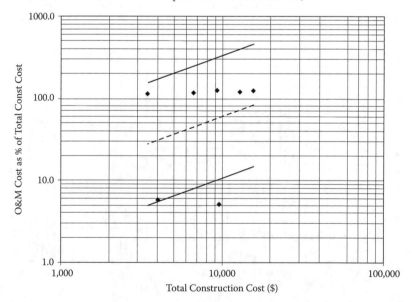

FIGURE 2.12
Annual O&M costs of infiltration trenches. (Data from SWRPC, 1991 Southeastern Wisconsin Regional Planning Commission. *Costs of urban nonpoint source water pollution control measures.* Waukesha: SWRPC; Landphair, H.C., J.A. McFalls, and D. Thompson, (2000). *Design methods, selections, and cost-effectiveness of stormwater quality structures.* Texas Transportation Institute, The Texas A&M University System, College Station, TX.)

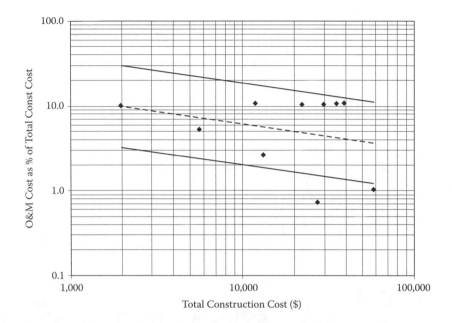

FIGURE 2.13
Annual O&M costs of bioinfiltration filters. (Data from Landphair, H.C., J.A. McFalls, and D. Thompson, (2000). *Design methods, selections, and cost-effectiveness of stormwater quality structures.* Texas Transportation Institute, The Texas A&M University System, College Station, TX; Wossink, A., and B. Hunt. 2003. *The economics of structural stormwater BMPs in North Carolina.* University of North Carolina Water Resources Research Institute report #UNC-WRRI-2003-344.)

operate the SMP. Rather than provide one estimate for total construction cost and another estimate for annual O&M expenditures, the data presented in the previous two sections have been combined to estimate the total present cost of each SMP as a function of size. As previously defined, the TPC is the sum of the total construction cost and the equivalent present cost of 20 years of annual O&M expenses. For each SMP, the TPC is estimated as a function of size (i.e., WQV or swale top width).

The total present cost with a 67% confidence interval for six of the seven SMPs was estimated as a function of water-quality volume. The total present cost of a 1000-ft long grassed/vegetative swale was estimated as a function of the swale top width. The TPC estimates incorporate the total construction cost data and annual O&M cost data presented in the previous sections. In this estimate, the annual O&M costs are converted to an equivalent present cost using historical data on the rates of municipal bond yields and inflation. The analysis method and the results for each of the seven SMPs are presented below.

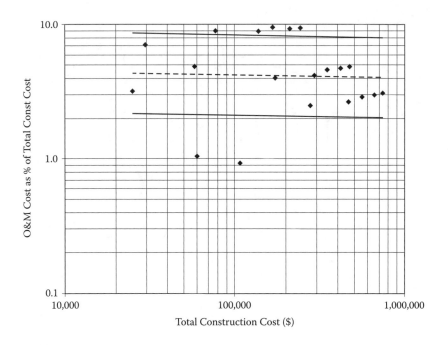

FIGURE 2.14
Annual O&M costs of sand filters. (Data from Landphair, H.C., J.A. McFalls, and D. Thompson, (2000). *Design methods, selections, and cost-effectiveness of stormwater quality structures.* Texas Transportation Institute, The Texas A&M University System, College Station, TX; Wossink, A., and B. Hunt. 2003. *The economics of structural stormwater BMPs in North Carolina.* University of North Carolina Water Resources Research Institute report #UNC-WRRI-2003-344.)

To estimate the TPC of each SMP the total construction cost was calculated as a function of size (i.e., WQV or swale top width) by multiplying the corresponding unit construction cost by WQV or, in the case of swales, by the swale length. Using these values of total construction cost and the annual O&M cost data best-fit line, the annual O&M cost was estimated for each WQV or swale top width. For example, for a dry detention basin, the unit construction costs for a range of WQVs were calculated from the best-fit line shown in Figure 2.2. The total construction costs were then estimated by multiplying the unit construction costs by the corresponding WQV. The annual O&M costs (as a percentage of construction cost) were then estimated using the best-fit line of Figure 2.9. Next, the value of the annual O&M cost estimates were calculated by multiplying each percentage (as found from the best-fit line) by the corresponding total construction cost. Finally, the annual O&M costs for a 20-year period were converted to an equivalent present cost (based on historical values of interest and inflation rates as described below) and added to the total construction cost.

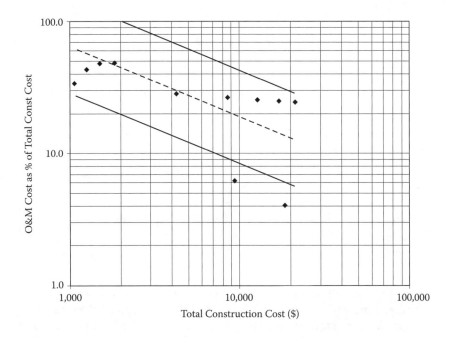

FIGURE 2.15
Annual O&M costs of grassed/vegetative swales. (Data from Landphair, H.C., J.A. McFalls, and D. Thompson, (2000). *Design methods, selections, and cost-effectiveness of stormwater quality structures.* Texas Transportation Institute, The Texas A&M University System, College Station, TX; SWRPC, 1991 Southeastern Wisconsin Regional Planning Commission. *Costs of urban nonpoint source water pollution control measures.* Waukesha: SWRPC.)

Before the conversion of the annual O&M costs to an equivalent present cost is described, it must be noted the annual O&M costs for infiltration trenches and grassed/vegetative swales were estimated in a different manner than described above. All but two of the O&M data points for these two SMPs (shown in Figures 2.12 and 2.15) were from Landphair, McFalls, and Thompson (2000) whose estimates ranged from 115 to 126% for infiltration trenches and 25 to 178% for grassed/vegetative swales. Because these values comprised most of the data and are high compared to the 5 to 20% for infiltration trenches and 5 to 7% for grassed/vegetative swales as summarized by the USEPA (1999), a different method was applied when estimating these annual O&M costs. For infiltration trenches and grassed/vegetative swales, average values of the annual O&M cost as a percentage of total construction cost as provided by the USEPA summary (Table 2.5) were assumed. Thus, annual O&M costs for infiltration trenches and grassed/vegetative swales were not determined from the best-fit line through the data, but rather assumed to be 12% (±7%) and 6% (±1%), respectively. Other than these assumptions, the TPC analysis for these two SMPs was identical to the others.

TABLE 2.6

Equation (2.4) Constants for Annual Operating and Maintenance Costs and Associated Confidence Intervals

Best Management Practice	Average Annual O&M Cost as % of TCC		Upper Confidence Interval		Lower Confidence Interval	
	β_0	β_1	β_0	β_1	β_0	β_1
Dry Basins	92.26	−0.327	97.50	−0.327	87.30	−0.327
Wet Basins	230.67	−0.362	328.85	−0.362	161.81	−0.362
Sand Filters	5.47	−0.023	10.89	−0.023	2.75	−0.023
Bioretention Filters	90.16	−0.292	273.53	−0.292	29.72	−0.292
Const. Wetlands	265.46	−0.396	584.79	−0.396	106.66	−0.396
Infiltration Trenches	0.081	0.718	0.423	0.718	0.0145	0.718
Grassed/Vegetated Swales	2,523.48	−0.531	5,688.53	−0.531	1,119.44	−0.531

Returning to the method used to convert the annual O&M costs to an equivalent present cost and having obtained an annual O&M cost estimate, it was assumed these costs would be incurred for 20 years. Based on this assumption, 20 years of annual O&M costs were converted to an equivalent present O&M cost using the time value of money and historical values of interest and inflation rates. Given an interest rate and inflation rate, the equivalent present cost of the 20-year annual O&M costs can be computed by an equation modified from Collier and Ledbetter (1988), which is:

$$P = C_{OM} \left[\frac{\left(\frac{1+r}{1+i}\right)^n - 1}{r - i} \right] \qquad (2.5)$$

where:

P = Equivalent present cost of 20-years of annual O&M costs
C_{OM} = annual O&M cost in the same year as P
r = inflation rate
i = interest rate
n = number of years (i.e., 20)

Equation (2.5) may be rewritten as

$$P = C_{OM} [E] \qquad (2.6)$$

where

$$E = \left[\frac{\left(\dfrac{1+r}{1+i}\right)^n - 1}{r - i} \right]$$

Using average annual AAA municipal bond yield rates (Mergent, Inc., 2003) for interest rate values and historical consumer price index (CPI) based inflation rates (Fintrend.com, 2004), the value of E was calculated for each year from 1944 through 2002. Because this analysis is based on a 20-year time span, the running 20-year average value of E was calculated for each year from 1963 through 2002. The running 20-year average values are shown in Table 2.7 and resulted in an overall average value of 18.68 +/−2.29 (67% confidence interval). Returning to the example and using a value of 18.68 for E, the present equivalent cost of 20 years of annual O&M expenses were calculated over the range of WQVs and added to the corresponding total construction cost to give the total present cost in 2005 dollars as a function of WQV. The uncertainties associated with the 67% confidence intervals of the unit construction costs, annual O&M costs as a percentage of the construction cost, and inflation and interest rates (i.e., E) were incorporated into the TPC as described by Kline (1985).

The total present cost (with 67% confidence interval), excluding land costs, of each SMP can be fit to an equation of the form:

TABLE 2.7

Yearly 20-year running average values of E.

Year	20-yr running Avg. E	Year	20-yr running Avg. E	Year	20-yr running Avg. E	Year	20-yr running Avg. E
1963	23.94	1973	17.55	1983	20.22	1993	18.23
1964	23.73	1974	18.25	1984	19.98	1994	17.41
1965	23.46	1975	18.68	1985	19.75	1995	16.91
1966	22.28	1976	18.74	1986	19.46	1996	16.74
1967	19.17	1977	18.82	1987	19.27	1997	16.36
1968	18.38	1978	19.02	1988	19.01	1998	15.93
1969	18.55	1979	19.80	1989	18.83	1999	15.12
1970	18.53	1980	20.56	1990	18.73	2000	14.32
1971	17.56	1981	20.66	1991	18.62	2001	14.12
1972	17.35	1982	20.46	1992	18.53	2002	14.27

TABLE 2.8

Equation (2.7) Constants for Total Present Cost and Associated Confidence Intervals

Best Management Practice	Average Total Present Cost β_0	β_1	Upper Confidence Interval β_0	β_1	Lower Confidence Interval β_0	β_1	Water Quality Volume Range (cubic ft)
Dry Basins	133.77	0.634	185.25	0.671	131.19	0.585	3000–3,560,000
Wet Basins	709.37	0.512	906.05	0.536	640.15	0.484	14,500–7,600,000
Sand Filters	740.96	0.594	1628.27	0.596	423.89	0.592	100–195,000
Bioretention Filters	97.08	0.776	291.86	0.723	51.47	0.802	900–35,000
Const. Wetlands	202.30	0.565	320.69	0.585	158.76	0.537	7,000–7,600,000
Infiltration Trenches	121.69	0.817	219.72	0.817	77.14	0.817	450–30,700

$$TPC = \beta_0 (WQV)^{\beta_1} \qquad (2.7)$$

where TPC = total present cost (2005 USEPA rainfall zone 1 dollars), WQV = water quality volume for which the SMP was designed to treat (ft^3), and β_0 and β_1 = constants. Values for β_0 and β_1 for each SMP are given in Table 2.8.

The constants presented in Table 2.8 are based on historical data and are intended to be used for comparative purposes only. They are not intended to estimate costs associated with specific SMPs nor should cost be the only factor considered when selecting a SMP.

Effectiveness of Contaminant Removal

An estimate of the total cost of a SMP can be a valuable aid during the planning and selection process. An inexpensive SMP that has minimal impact on water quality, however, would be of little value. Thus, knowledge of the impact or effectiveness a particular SMP will have on water quality is just as important as the cost. In an effort to provide information in this area, an analysis was performed in which the total amount of total suspended solids (TSS) and phosphorus removed over a 20-year span was estimated as a function of water quality volume. In this analysis, the amount of TSS and P removed is considered a function of the fraction of stormwater runoff that will be treated by the SMP over its 20-year life, the pollutant load that reaches the SMP, and the removal performance of the SMP itself. Of course, some of the variables listed above depend on other variables such as watershed area, impervious area, rainfall amounts, and so on. All of these variables and the analytical method that was used to incorporate them into the estimate of total pollutant load removal are described and discussed below.

Runoff Fraction Treated

Most SMPs are designed for a particular rainfall depth used to estimate a water-quality volume or, in the case of swales, filter strips, and similar SMPs, a peak flow rate. The WQV or peak flow rate is used to size the SMP. An SMP is designed for a finite value of rainfall or runoff, thus there is always the chance that a given storm will produce more runoff than the unit was designed to store or treat. When that happens, a portion of the runoff bypasses the SMP or is discharged from the SMP via an overflow outlet and receives no treatment. To account for this untreated fraction of the runoff, a statistical analysis was performed on historical rainfall data in the Minneapolis–St. Paul metropolitan region (i.e., the Twin Cities). The values from the Twin Cities will approximate most Midwestern regions. For other regions the process, as described below, can be used to estimate the fraction of stormwater runoff that will be bypassed or exit the SMP without treatment for a given rainfall depth.

Inasmuch as design recommendations for SMPs usually state the devices should be designed to drain in two days, two-day running sum precipitation amounts in the Twin Cities were calculated and analyzed from 1950 through 2003. For example, if the precipitation depths measured on four consecutive days were 0.21 in., 0.13 in., 0.35 in., and 0.07 in., the data would be combined into two-day precipitation amounts of 0.34 in., 0.48 in., and 0.42 in., respectively. Using the combined data, a two-day running sum (R_S) histogram was generated using 0.10-in. increments from zero to four inches, with the last bin including any two-day sum that was greater than or equal to four inches. Of the 9,720 nonzero entries, five fell into the latter category, with the largest having a value of 10.00 in. Columns 1 through 4 of Table 2.9 show the histogram in tabular form along with the frequency and cumulative frequency distributions. Subtracting the cumulative frequency from 1.00 and multiplying by 100 gives the percent exceedance as shown in column 5 and plotted in Figure 2.16.

Thus, Table 2.9 or Figure 2.16 can be used to determine the fraction of two-day precipitation events that exceeded a particular precipitation depth. For example, based on Figure 2.16, a two-day rainfall depth of 1.00 in. was exceeded approximately 7% of the time over the 54-year period analyzed. Alternatively, using Table 2.9 and linearly interpolating between

TABLE 2.9

Statistical Analysis of Historical Two-Day Precipitation Amounts at the Minneapolis-St. Paul Airport

(1) Range (in.)	(2) # of events	(3) Frequency	(4) Culm. Frequency	(5) % Exceedance	(6) Area (in.)	(7) Culm. Area (in.)	(8) % of Total Area	(9) Rainfall Depth (in.)
				100.00			0.00	0.00
0<Rs<0.1	4037	0.41533	0.41533	58.47	3.962	3.962	13.88	0.05
0.1<=Rs<0.2	1599	0.16451	0.57984	42.02	5.024	8.986	31.48	0.15
0.2<=Rs<0.3	965	0.09928	0.67912	32.09	3.705	12.691	44.45	0.25
0.3<=Rs<0.4	683	0.07027	0.74938	25.06	2.858	15.549	54.46	0.35
0.4<=Rs<0.5	501	0.05154	0.80093	19.91	2.248	17.797	62.34	0.45
0.5<=Rs<0.6	377	0.03879	0.83971	16.03	1.797	19.594	68.63	0.55
0.6<=Rs<0.7	276	0.02840	0.86811	13.19	1.461	21.055	73.75	0.65
0.7<=Rs<0.8	250	0.02572	0.89383	10.62	1.190	22.245	77.92	0.75
0.8<=Rs<0.9	171	0.01759	0.91142	8.86	0.974	23.219	81.33	0.85
0.9<=Rs<1.0	139	0.01430	0.92572	7.43	0.814	24.033	84.18	0.95
1.0<=Rs<1.1	115	0.01183	0.93755	6.24	0.684	24.717	86.58	1.05
1.1<=Rs<1.2	99	0.01019	0.94774	5.23	0.574	25.290	88.59	1.15
1.2<=Rs<1.3	98	0.01008	0.95782	4.22	0.472	25.763	90.24	1.25
1.3<=Rs<1.4	65	0.00669	0.96451	3.55	0.388	26.151	91.60	1.35
1.4<=Rs<1.5	58	0.00597	0.97047	2.95	0.325	26.476	92.74	1.45
1.5<=Rs<1.6	46	0.00473	0.97521	2.48	0.272	26.748	93.69	1.55
1.6<=Rs<1.7	34	0.00350	0.97870	2.13	0.230	26.978	94.50	1.65
1.7<=Rs<1.8	27	0.00278	0.98148	1.85	0.199	27.177	95.20	1.75
1.8<=Rs<1.9	20	0.00206	0.98354	1.65	0.175	27.352	95.81	1.85

1.9<=Rs<2.0	18	0.00185	0.98539	1.46	0.155	27.507	96.35	1.95
2.0<=Rs<2.1	18	0.00185	0.98724	1.28	0.137	27.644	96.83	2.05
2.1<=Rs<2.2	16	0.00165	0.98889	1.11	0.119	27.764	97.25	2.15
2.2<=Rs<2.3	14	0.00144	0.99033	0.97	0.104	27.868	97.62	2.25
2.3<=Rs<2.4	16	0.00165	0.99198	0.80	0.088	27.956	97.93	2.35
2.4<=Rs<2.5	9	0.00093	0.99290	0.71	0.076	28.032	98.19	2.45
2.5<=Rs<2.6	9	0.00093	0.99383	0.62	0.066	28.098	98.42	2.55
2.6<=Rs<2.7	10	0.00103	0.99486	0.51	0.057	28.155	98.62	2.65
2.7<=Rs<2.8	8	0.00082	0.99568	0.43	0.047	28.202	98.79	2.75
2.8<=Rs<2.9	8	0.00082	0.99650	0.35	0.039	28.241	98.92	2.85
2.9<=Rs<3.0	8	0.00082	0.99733	0.27	0.031	28.272	99.03	2.95
3.0<=Rs<3.1	5	0.00051	0.99784	0.22	0.024	28.296	99.12	3.05
3.1<=Rs<3.2	5	0.00051	0.99835	0.16	0.019	28.315	99.18	3.15
3.2<=Rs<3.3	2	0.00021	0.99856	0.14	0.015	28.331	99.24	3.25
3.3<=Rs<3.4	2	0.00021	0.99877	0.12	0.013	28.344	99.28	3.35
3.4<=Rs<3.5	2	0.00021	0.99897	0.10	0.011	28.355	99.32	3.45
3.5<=Rs<3.6	0	0.00000	0.99897	0.10	0.010	28.365	99.36	3.55
3.6<=Rs<3.7	2	0.00021	0.99918	0.08	0.009	28.375	99.39	3.65
3.7<=Rs<3.8	2	0.00021	0.99938	0.06	0.007	28.382	99.42	3.75
3.8<=Rs<3.9	1	0.00010	0.99949	0.05	0.006	28.388	99.44	3.85
3.9<=Rs<4.0	0	0.00000	0.99949	0.05	0.005	28.393	99.45	3.95
4.0<=Rs	5	0.00051	1.00000	0.00	0.156	28.548	100.00	10.00
Note: # of Events = 9720				Total Area =	0.156	28.548		

28.548

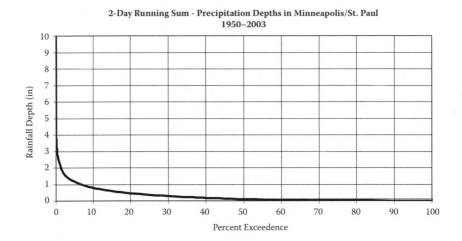

FIGURE 2.16
Exceedance probabilities of two-day precipitation depths in the Twin Cities.

7.43 and 6.24% gives a value of 6.84% exceedance for a precipitation depth of 1.00 in. Furthermore, if an SMP were designed for a precipitation depth of 1.00 in., the graph area that is both under the curve and below the horizontal line that corresponds to an abscissa value of 1.00 in. divided by the total area under the curve, equals the fraction of the two-day summed precipitation amounts that were below the 1.00 in. design storm depth. The values of the graph area, cumulative area, and percentage of total area corresponding to each precipitation depth have been calculated and are shown in columns 6, 7, and 8, respectively, of Table 2.9.

Due to infiltration and other abstractions of the stormwater, which occur as the runoff makes its way to the SMP, this ratio is not exactly the fraction of runoff that would be treated by the SMP. That would only be the case if 100% of the precipitation were to reach the SMP as runoff. The percentage of total area values shown in column 8, however, can be used as an approximate estimate of the fraction of runoff that would be treated by an SMP designed for the corresponding rainfall depth in column 9. For example, based on columns 8 and 9 of Table 2.9, if an SMP were designed for a precipitation depth of 2.25 inches, it could be estimated that, based on historical data, the SMP would treat 97.62% of the stormwater runoff over time.

For the purposes of this analysis, it was assumed that all SMPs would be designed for a precipitation depth of 1.45 in., which is approximately the three-month, 24-hour storm for the Twin Cities metro region (Huff and Angel, 1992). As shown in Table 2.9, a depth of 1.45 in. corresponds to approximately 93% of the area under the curve. As discussed above, it was estimated that 93% of all stormwater runoff will be treated by an

SMP, which is designed for the volume of runoff corresponding to this precipitation depth. Thus, when estimating the total amount of TSS and P removed over 20 years, it was assumed that 93% of all stormwater runoff will be treated by the SMPs and the remaining 7% of the runoff will receive no treatment. Thus, total suspended sediment and phosphorus removal are given by:

$$\%\text{Total Removal} = 0.93 * (\%\text{Removal by SMP}) \qquad (2.8)$$

where the "%Removal by SMP" is the removal based on inflow and treated outflow concentrations and does not consider overflow conditions. Overflow or bypass conditions are accounted for by multiplying the "%Removal by SMP" by 0.93.

Pollutant Loading

Several methods with a wide degree of complexity are available to estimate stormwater pollutant loads. For example, the Stormwater Management Model (SWMM) is public domain software and can be used to model single storm events or watershed basins over time. Additional methods described by Young et al. (1996) include regional United States Geological Survey (USGS) equations for estimation of storm loads, runoff volumes, and event mean concentrations. These equations have been developed for three regions in the United States and are based on regression analysis of nationwide data. A simplified, albeit less accurate, set of USGS regression equations is available and can be used to estimate storm runoff loads and volumes. The USGS has derived a set of equations to estimate storm mean concentrations and mean seasonal or annual loads.

The Federal Highway Administration (FHWA) has developed a method to estimate pollutant loading from highway runoff (Driscoll, Shelley, and Strecker. 1990). As with the USGS methods, the FHWA method is a regional method, in this case with nine regions, which involves a relatively large amount of detailed input to arrive at an estimate of annual pollutant mass loading.

The methods described above require a level of detail that is well beyond what is necessary (or perhaps possible) for the comparative purposes of this analysis. Thus, a modified version of a less involved method, the simple method, was selected to estimate pollutant loads. The simple method, first proposed by Schueler (1987), is widely accepted and requires only the mean annual precipitation, a percentage of rainfall events that produce no runoff, the drainage area, and a runoff coefficient be known. The previously mentioned modified simple method is used by the Lower Colorado River Authority (1998) and has been recommended for use by the State of Texas, Department of Transportation (Landphair et al., 2000). The modified simple method is:

$$L = (0.2266) * A * R_F * R_V * C \qquad (2.9)$$

where:
L = Annual pollutant load (lb.)
A = Watershed area (acre)
R_F = Average annual rainfall (in.)
R_V = Average annual runoff coefficient (i.e., runoff: rainfall ratio)
C = Average annual contaminant (i.e., TSS & P) concentration (mg/L)

The runoff coefficient R_V, which was described for water quality volume calculations previously, is estimated as

$$R_V = 0.05 + 0.009 * (I) \qquad (2.2)$$

where I = percentage of watershed that is impervious.

To coincide with the 20-year time span used to estimate the total present cost, the pollutant loading must be estimated for 20 years. To accomplish this, Equation (2.9) must be multiplied by 20. The variable R_F is no longer to be defined as the average annual rainfall but rather the 20-year running average of annual rainfall (inches). Incorporating these small but significant changes, the equation used to estimate the TSS and P loading over a 20-year span becomes:

$$L_{20} = 20 * 0.2266 * A * R_{F20} * R_V * C \qquad (2.10)$$

where:
L_{20} = Estimated pollutant load over 20 years (lb)
R_{F20} = 20-year running average of annual rainfall (in.)

In addition, all other variables are as previously defined.

For the purposes of this analysis it was assumed the watershed area A, percentage impervious I, and therefore the runoff coefficient R_V, would be known without any uncertainty. To obtain an estimate of R_{F20}, a statistical analysis on historical precipitation data in Minneapolis and St. Paul from 1950 through 2003 was performed. The results showed the 20-year running average precipitation depth is 28.44 inches +/- 1.80 inches (67% confidence interval).

To determine estimates of the average annual concentration of TSS and P in stormwater runoff, data were compiled from several studies and dozens of sites (Moxness, 1986, 1987, 1988; Driscoll et al., 1990; Oberts, 1994; Barrett et al., 1995; Stanley, 1996; Wu, Holman, and Dorney, 1996; Sansalone and Buchberger, 1997; Barrett, et al., 1998; Anderle, T.A., 1999; Legret and Colandini, 1999; Waschbusch, Selbig, and Bannerman, 1999; Carleton et al., 2000; Drapper, Tomlinson, and Williams, 2000; Brezonik and Stadelmann, 2002; Harper, Herr, and Livingston, undated). Data analysis revealed the average values of stormwater concentrations of TSS and P from sites located in the Twin Cities were essentially the same as average values of all other sites located throughout the nation and Australia. Because the data were similar, the national average values of 131 mg/L +/− 77 mg/L (67% confidence interval) for TSS and 0.55 mg/L +/− 0.41 mg/L (67% confidence interval) for total P were used. With values for R_{F20} and C estimated, the total pollutant load for TSS and P in pounds over a 20-year time frame, as estimated by Equation (2.10), becomes a function of only two variables: watershed area and, with the use of Equation (2.2), the percentage of the watershed area that is impervious.

With the selection of a storm design depth of 1.45 in. as previously discussed, the two remaining variables that determine the 20-year pollutant loads (i.e., watershed area and percentage impervious) are the same two variables that determine the WQV. Thus, for a watershed of known area and percentage impervious, both the WQV and the TSS and P loads over 20 years can be estimated. In other words, for a given watershed, each value of WQV corresponds to a unique value of 20-year TSS and P loads. Although pollutant loading is certainly important, the intent of this analysis is to estimate the load removed by the SMPs over a 20-year span. As with the total present cost, the estimate of the pollutant load removed by each SMP is estimated as a function of WQV. Before this analysis can be completed, however, one remaining variable, the percentage of TSS and P removed by each category of SMPs, must be estimated.

Fraction of Contaminants Removed

With the fraction of runoff treated and the total 20-year pollutant load estimated, the remaining variable that must be estimated is the fraction of TSS and P removed by each type of SMP (i.e., "%Removal by SMP" in Equation (2.5)). Once the removal rate of each SMP has been estimated, the total mass of TSS and P removed over the 20-year span may be estimated by multiplying the 20-year pollutant load by both the fraction of runoff treated (i.e., estimated to be 93% for a design precipitation of 1.45 in.) and the fraction of pollutant removed by the SMP. To make the estimate of SMP removal performance (i.e., %Removal SMP in Equation (2.8)) as realistic as possible, published data on the performance of the various types of SMPs analyzed were collected and the average removal rate with 67% confidence interval calculated.

Only data from actual sites that underwent field testing were included. When a single site was monitored over time and had more than one removal rate reported, only the average value of the data for that site was included in the analysis. Most of the published data were reported as a percentage of change between influent and effluent concentrations, although a few considered the percent change between the total mass load entering the SMP and the mass load exiting the SMP. These two techniques will give different results, but with the large uncertainty in performance data, the techniques used did not seem to add uncertainty to the results. For each type of SMP the average percentage removal of the combined data was calculated and assumed the average percentage of mass load removed. The infiltration of stormwater, which may occur inside some SMPs (e.g., wetlands, dry basins, etc.), was not considered in the concentration-based performance calculations. The percentage drop in the influent to effluent concentration should be smaller than the percentage of mass load removed. Thus, by combining concentration-based removal rates with those based on mass loads and assuming the resulting average to be the percentage of mass load removed, the removal rate estimate is conservative.

The results are given in Table 2.9. Sufficient amounts of reliable data that are needed to estimate the TSS removal rate of bioretention filters and TSS and phosphorus removal rates of infiltration trenches were not available. As denoted by the asterisks in Table 2.9, typical values of 90 and 75% for TSS removal (for bioretention filters and infiltration trenches, respectively) as reported by the *Idaho BMP Manual* (undated), were used. Assumed was the *Idaho BMP Manual* typical infiltration trench phosphorus removal of 55%. The assumed values for TSS removal were either in agreement with other reported typical ranges of effectiveness, or conservative (Caltrans, 2004; MPCA, 2000). The 67% confidence intervals for these SMPs were assumed and are denoted by an asterisk in Table 2.10.

As previously discussed, the published data used to calculate the values shown in Table 2.10 were reported as either percentage drop in concentration between influent and effluent stormwater or percentage removal of the total mass load entering the SMP. The values are based only on stormwater treated by the SMP and do not account for any portion of the flow that bypasses the SMP or exits through an overflow outlet. The confidence intervals reported in Table 2.10 reveal a large amount of uncertainty in the reported data. The uncertainty is likely due to variations in design and maintenance of the SMPs. If proper maintenance is not performed, removal levels will drop. Parameters such as swale slope, pond, and wetland residence time, and so on, affect removal.

The average total amount of TSS and phosphorus that can be expected to be removed by each SMP (except for grassed/vegetative swales) was calculated by multiplying the average 20-year total contaminant load

TABLE 2.10

Average Percent Removal Rates of SMPs with Corresponding
Confidence Interval

SMP	% TSS Removal	TSS 67% CI	% P Removal	P 67% CI
Dry Detention Basins	53	±28	25	±15
Wet Basins	65	±32	52	±23
Stormwater Wetland	68	±25	42	±26
Bioretention Filter	90*	±10*	72	±11
Sand Filter	82	±14	46	±21
Infiltration Trench	75*	±10*	55*	±35*
Filter Strips/Grassed Swales	75	±20	41	±33

by 93% (i.e., the estimated percent runoff treated) and the corresponding removal rate as shown in Table 2.9. Confidence intervals of 67% were
estimated by incorporating uncertainty in stormwater contaminant concentrations, percentage contaminant removed by the SMP, and 20-year
precipitation amounts by the direct analytical method described by
Kline (1985).

Because the weight of contaminant removed is directly proportional to
the volume of water treated, the amount of contaminant expected to be
removed over a 20-year design life can be fit to an equation of the form:

$$C_R = \beta_0(WQV) \tag{2.11}$$

where C_R = weight of contaminant removed over a 20-year design life (lbs),
WQV = design water quality volume (ft³), and β_0 = constant. The values
of β_0 for TSS and phosphorus for each SMP for the average weight of contaminant expected to be removed and the corresponding 67% confidence
intervals are given in Table 2.11.

Because swales are designed for a peak flow rate and not WQV, an estimate of the total load removed by swales over 20 years could not be estimated as a function of WQV. The volume of runoff that will be treated by a
swale can be estimated, however, the removal rates reported in Table 2.10
for Filter Strips/Grassed Swales may be used to estimate the corresponding total contaminant load removed.

Example

SMPs under consideration for a 50-acre watershed that is 80% impervious
include a dry detention basin and a surface sand filter. The SMP is to be

TABLE 2.11

Equation (2.11) Constants for Contaminant Removal over 20-Year Design Life for Average Values and Associated 67% Confidence Intervals

Stormwater Management Practice	TSS			Phosphorus		
	Average	Upper Confidence Interval	Lower Confidence Interval	Average	Upper Confidence Interval	Lower Confidence Interval
Dry Basins	1.718	2.840	0.597	0.0032	0.0059	0.0004
Wet Basins	2.104	3.413	0.796	0.0065	0.0110	0.0014
Sand Filters	2.650	3.748	1.553	0.0057	0.0103	0.0012
Bioretention Filters	2.923	4.034	1.812	0.0112	0.0185	0.0039
Const. Wetlands	2.204	3.359	1.050	0.0052	0.0099	0.0005
Infiltration Trenches	2.436	3.408	1.464	0.0069	0.0131	0.0007

designed for a 1.45-in. precipitation depth and a comparison of the cost and effectiveness of both SMPs is desired.

Using Equations (2.1) and (2.2), the WQV can be determined as follows:

$$WQV = \left(\frac{43560}{12}\right) * 1.45 * \left(0.05 + 0.009(80)\right) * 50$$

$$WQV \approx 200,000 \text{ ft}^3$$

From Equation (2.7) and Table 2.8 the TPC of an average dry detention basin of this size is just over $300,000 with a 67% confidence interval range of about $170,000 to $675,000. A similarly sized average surface sand filter would, based on Equation (2.7) and Table 2.8 cost approximately $1,000,000 with a 67% confidence interval range of $600,000 to $2,300,000. For a comparison among all SMPs, Table 2.12 lists the estimated average TPC of all practices analyzed herein for various WQVs. For each SMP, TPCs are not estimated for WQVs that are outside the range of the original construction cost data. Thus some values in Table 2.12 do not have a cost entry.

Investigation of Table 2.12 reveals that, based on the collected data and in terms of TPC, wetlands are the least expensive SMP for the range of WQVs listed. This finding is somewhat similar to that of Wossink and Hunt (2003) who concluded that, in terms of construction costs, wetlands were the least expensive of four SMPs (wet ponds, constructed wetlands, sand filters, bioretention basins) for watersheds larger than 10 acres in sandy soils. Contrary to the previous conclusions, the California

TABLE 2.12

Average Total Present Cost (in $1,000s) of SMPs at varying WQVs[a]

SMP	Water Quality Volume (ft³)				
	3,000	10,000	30,000	100,000	250,000
Dry Det. Basin	22	46	91	198	359
Wet/Ret. Basin	47	83	141	256	407
Const. Wetland	21	38	68	131	219
Infilt. Trench	84	226	554	—	—
Bioinfilt. Filter	49	122	286	—	—
Sand Filter	86	176	338	691	—

[a] Land costs are excluded, and need to be determined separately.

Stormwater Quality Association (2003) states that wetlands are relatively inexpensive but are typically 25% more expensive than stormwater ponds of equivalent volume. Furthermore, due to an abundance of phosphorus accumulated in a wetland and low dissolved oxygen concentrations in the water within the wetland, instances have been reported in which a wetland becomes a phosphorus source instead of a sink. Removing vegetation from the wetland in an attempt to lower phosphorus levels may seem like a viable option, however, Kadlec and Knight (1996) report that this has a minimal impact on wetland phosphorus levels. One must remember that inasmuch as wetlands generally require more land area, any savings in TPC may potentially be more than offset by larger land acquisition costs.

Returning to the example, over 20 years of the estimated TSS removal and 67% confidence interval for the dry detention basin can, with the use of Equation (2.11) and Table 2.11 be estimated to be 344,000 pounds with a range of 120,000 pounds to 570,000 pounds. The corresponding sand filter TSS removal based on Equation (2.11) and Table 2.11 is estimated to be 550,000 pounds with a range from 310,000 pounds to 750,000 pounds.

The phosphorus removed over 20 years can be estimated in a similar manner using Equation (2.11) and Table 2.11. For the dry detention basin the average P removal is approximately 630 pounds with a range of 80 to 1,200 pounds (67% confidence interval). The sand filter average P removal is about 1,100 pounds with a range from about 220 pounds to about 2,000 pounds. Thus, for this watershed and design depth, the dry detention pond, on average, would cost less to construct (not including land costs) but it would remove less TSS and phosphorus. However, land costs must always be considered.

Focusing on associated land costs of each SMP under consideration, Table 2.3 can be used to estimate the range of expected land area required for each SMP. Using the values based on total watershed area and selecting the high end of each range, the dry detention basin would require 2.0% of the total watershed area resulting in a basin land area of 1 acre. Similarly, the sand filter would require 3.0% of the impervious area (i.e., 40 acres) or 1.2 acres. If land costs are known, the land areas can be used to estimate land costs associated with each SMP. For example, if land costs were $10,000 per acre, acquiring the land for the detention basin would cost an additional $10,000 and the land for the sand filter would cost $12,000. The resulting total cost (now including a rough estimate for land acquisition) for the detention basin and surface sand filter would be $310,000 and $1,012,000, respectively. Thus, the dry detention pond would still be cheaper (but less effective), on average.

The sand filter is estimated to remove more TSS and phosphorus, meaning the parties involved would have to weigh the increased cost of the sand filter against its added benefit (i.e., more contaminant removal). This example and the intended use of the information contained in this chapter are preliminary in nature; to obtain a more accurate estimate of costs a more detailed design of each SMP should be completed.

References

ASCE. (2004). International stormwater best management practices (SBMP) database. American Society of Civil Engineers, http://www.SBMPdatabase.org/

Anderle, T.A. (1999). Analysis of stormwater runoff and lake water quality for the Twin Cities metropolitan area. M.S. Thesis, University of Minnesota.

Barrett, M., J. Malina, R. Charbeneau, and G. Ward. (1995). *Characterization of highway runoff in the Austin, TX area.* Center for Research in Water Resources, UT-Austin, Austin, TX.

Barrett, M., P. Walsh, J. Malina, and R. Charbeneau. (1998). Performance of vegetative controls for treating highway runoff. *Jour. Environ. Engrg.*, 124: 1121–1128.

Brezonik, P., and T. Stadelmann. (2002). Analysis and predictive models of stormwater runoff volumes, loads, and pollutant concentrations from watersheds in the Twin Cities metropolitan area, Minnesota. *Water Res.*, 36: 1743–1757.

Brown, W., and T. Schueler. (1997). *The economics of stormwater BMPs in the mid-Atlantic region: Final report.* Center for Watershed Protection, Silver Spring, MD.

California Stormwater Quality Association. (2003). *California stormwater BMP handbook – New development and redevelopment.* Menlo Park, CA, USA. Available online at http://www.cabmphandbooks.com/Documents/Development/TC-21.pdf

Caltrans. (2004). *BMP retrofit pilot program – Final report, Appendix C3.* California Department of Transportation, Division of Environmental Analysis, Sacramento, CA. Available at http://www.dot.ca.gov/hq/env/stormwater/special/newsetup/_pdfs/new_technology/CTSW-RT-01-050.pdf

Carleton, J., T. Grizzard, A. Godrej, H. Post, L. Lampe, and P. Kenel. (2000). Performance of a constructed wetlands in treating urban stormwater runoff. *Water Environ. Res.*, 72(3): 295–304.

Claytor, R.A., and T.R. Schueler. (1996). *Design of stormwater filtering systems.* Center for Watershed Protection, Silver Spring, MD.

Collier, C.A., and W.B. Ledbetter. (1988). *Engineering Economic and Cost Analysis,* 2nd ed. New York: Harper & Row.

Drapper, D., R. Tomlinson, and P. Williams. (2000). Pollutant concentrations in road runoff: Southeast Queensland case study. *J. Environ. Engrg.,* 126(4): 313–320.

Driscoll, E., P.E. Shelley, and E.W. Strecker. (1990). *Pollutant loadings and impacts from highway stormwater runoff, Volumes I–IV.* United States Department of Transportation, Federal Highway Administration, Report No. FHWA/RD-88-006-9.

Fintrend.com. (2004). InflationData.com. Financial trend forcaster, http://inflationdata.com/inflation/Inflation_Rate/HistoricalInflation.aspx

Harper, H.H., J.L. Herr, and E.H. Livingston. (Undated). Alum treatment of stormwater runoff—An innovative BMP for urban runoff problems. Unpublished. Environmental Research and Design, Inc., Orlando, FL and Florida Department of Environmental Protection, Tallahassee, FL.

Huff, F.A., and J.R. Angel. (1992). *Rainfall frequency atlas of the Midwest.* State Water Survey Division, State of Illinois, Champaign, IL.

Idaho Department of Environmental Quality. (Undated). *BMP Manual.* Boise, ID.

Kadlec, R.H., and R.L. Knight. (1996). *Treatment Wetlands.* Boca Raton, FL: Lewis.

Kang, J.H., P.T. Weiss, J.S. Gulliver, and B.C. Wilson. (2008). Maintenance of stormwater BMPs: Frequency, effort, cost. *Stormwater,* 9(8, Nov.–Dec.): 18–28.

Kline, S.J. (1985). The purposes of uncertainty analysis. *J. Fluids Engrg.,* 107: 153–160.

Landphair, H.C., J.A. McFalls, and D. Thompson, (2000). *Design methods, selections, and cost-effectiveness of stormwater quality structures.* Texas Transportation Institute, The Texas A&M University System, College Station, TX.

Legret, M., and V. Colandini. (1999). Effects of a porous pavement with reservoir structure on runoff water: Water quality and fate of heavy metals. *Water Sci. Tech.,* 39(2): 111–117.

Lower Colorado River Authority. (1998). *Nonpoint Source Pollution Control Technical Manual,* 3rd ed. Austin, TX: LCRA.

Mergent, Inc. (2003). *Mergent Municipal & Government Manual.* New York: Mergent.

Minnesota Pollution Control Agency (MPCA) (2000). *Protecting water quality in urban areas-best management practices for dealing with stormwater runoff from urban, suburban, and developing areas of Minnesota.* St. Paul.

Moxness, K. (1986). *Characteristics of urban freeway runoff, phase I.* Water Quality Unit, Environmental Services Section, Office of Technical Support, St. Paul: Minnesota Department of Transportation.

Moxness, K. (1987). *Characteristics of urban freeway runoff, phase II.* Water Quality Unit, Environmental Services Section, Office of Technical Support, St. Paul: Minnesota Department of Transportation.

Moxness, K. (1988). *Characteristics of urban freeway runoff, phase III.* Water Quality Unit, Environmental Services Section, Office of Technical Support, St. Paul: Minnesota Department of Transportation.

Oberts, G. (1994). Influence of snowmelt dynamics on stormwater runoff quality. *Watershed protection techniques* 1(2), Elliott City, MD: Center for Watershed Protection.

Pitt, R.E., and J.G. Voorhees. (1997). Storm water quality management through the use of detention basins, a short course on storm water detention basin design basics by integrating water quality with drainage objectives. April 29–30 and May 21–22, St. Paul: University of Minnesota.

Sansalone, J.J., and S.G. Buchberger. (1997). Partitioning and first flush of metals in urban roadway stormwater. *J. Environ. Engrg.*, 123: 134.

Schueler, T.R. (1987). *Controlling urban runoff: A practical manual for planning and designing urban BMPs.* Department of Environmental Programs, Washington, DC: Metropolitan Council of Governments.

SWRPC (Southeastern Wisconsin Regional Planning Commission.) (1991). *Costs of urban nonpoint source water pollution control measures.* Waukesha, WI.

Stanley, D. (1996). Pollutant removal by a stormwater dry detention pond. *Water Environ. Res.*, 68(6): 1076–1083.

Turner Construction. (2004). Building cost index, 2004 fourth quarter forecast. New York. http://www.turnerconstruction.com/corporate/content.asp?d=20

Urban Drainage Flood Control District (UCFCD), Denver CO. (1992). *Best Management Practices, Urban Storm Drainage Criteria Manual*, Vol. 3, Denver.

USEPA. (1999). *Preliminary data summary of urban stormwater best management practices.* EPA-821-R-99-012, Washington, DC.

Waschbusch, R., W. Selbig, and R. Bannerman. (1999). *Sources of Phosphorus in Stormwater and Street Dirt from Two Urban Residential Basins in Madison, WI, 1994-95.* USGS Water-Resources Investigations Report 99-4021.

Wossink, A., and B. Hunt. (2003). *The economics of structural stormwater BMPs in North Carolina.* University of North Carolina Water Resources Research Institute report #UNC-WRRI-2003-344.

Wu, J.S., R.E. Holman, and J.R. Dorney. (1996). Systematic evaluation of pollutant removal by urban wet detention ponds. *J. Environ. Engrg.*, 122(11): 983–988.

Young, G.K., S. Stein, P. Cole, T. Kammer, F. Graziano, and F. Bank. 1996. *Evaluation and management of highway runoff water quality.* United States Department of Transportation, Federal Highway Administration, Pub. No. FHWA-PD-96-032. Washington, DC.

3

Economic Costs, Benefits, and Achievability of Low-Impact Development-Based Stormwater Regulations

John B. Braden and Amy W. Ando

CONTENTS

Introduction

Converting land from natural to developed status causes discharges of sediment and other pollutants during construction. In addition, the impervious surfaces and grading alterations accompanying development cause ongoing flows of environmental damage in the form of reductions in groundwater recharge and increased variability of stormwater runoff leading to stream destabilization and increased flood risks. Beginning in 2000, the U.S. Environmental Protection Agency (EPA) was placed under court order to promulgate effluent guidelines for stormwater discharges from real estate development and construction (C&D) projects under the Clean Water Act.[*] A growing literature documents the ability of low-impact style development (LID) measures to reduce water pollution, improve hydrological function, and save developers money (USEPA, 2007). Nevertheless, stormwater regulations that would effectively mandate such development techniques met with stiff opposition from the home building industry (e.g., Crowe, 2002). To inform the policy debate over promoting LID and enhanced stormwater control, this chapter assesses the economic costs and benefits of such regulations, including whether they have the potential to adversely affect housing markets and employment. Our assessment of published evidence suggests that LID measures can achieve stormwater pollution reductions at a cost less than is typically estimated for conventional "end-of-pipe" strategies. In addition, LID technologies last beyond the construction phase of development and offer substantial complementary environmental benefits, in addition to mitigating pollution.

Real estate development can cause various negative externalities related to stormwater flows and hydrological change. An externality exists when the actions of one individual affect the interests of another, either positively or negatively. For construction, vegetative cover is removed in favor of impervious roofs, streets, driveways, and parking lots. During the construction and development phase, disrupting vegetation and soils typically leads to large increases in erosion, sedimentation, and delivery of dissolved and suspended pollutants to surface waters. In the longer term, manicured landscapes and impervious structures concentrate and accelerate stormwater flows while diminishing areas available for infiltration. Heightened erosion, siltation, and discharge of landscape chemicals and debris persist while, in addition, flooding is exacerbated downslope and downstream. Reduced infiltration starves aquifers of replenishment (USEPA, 1997). The resulting impacts may be relatively small at the scale of individual parcels, but they are large cumulatively when considering the more than 2 million acres developed each year in the continental United States (USDA, 2007).

The most direct approach to resolve stormwater problems would be for all affected parties to agree on solutions. However, private negotiations are usually impractical because stormwater impacts are diffuse and cumulative in time and space. A web of regulations and programs has thus emerged to try to control problems associated with stormwater runoff and development.[†] As part of the National Pollution Discharge Elimination System (NPDES) under the Clean Water Act, municipalities and industrial

---

[*] Documents associated with ongoing rulemaking (USEPA, 2008a,b,c, and d) are available at www.epa.gov/waterscience/guide/construction (accessed October 27, 2009). The statutory and regulatory history is outlined in USEPA (2008a).

[†] See http://www.stormwaterauthority.org/regulatory_data/ for detailed information about those regulations.

dischargers must comply with stormwater permitting requirements and developers must control sediment production during construction and development (with an emphasis on sediment capture). In 1990, these "Phase I" requirements were applied to construction sites larger than five acres and municipalities with populations of 100,000 or more; in 1999, the "Phase II" program extended Phase I permitting requirements to cover small urbanized municipalities and construction sites larger than one acre. The particular NPDES stormwater requirements vary by state, often requiring developers and cities to provide for the collection, detention, and conveyance of runoff during the construction phase of development. However, to date, NPDES permits have little to say about the hydrological effects of altered landscapes or the quality of runoff following the construction process. There is concern that the sediment capture requirements for construction sites are often ineffectual (Line and White, 2001; Hayes, McLaughlin, and Osmond, 2005). Thus, externalities persist due to gaps in local, state, and federal regulations of stormwater.

Policies to control externalities can be promulgated at various levels of government. The literature on environmental federalism (e.g., Braden and Proost, 1997; Oates, 2001) indicates that an externality problem may be best handled by a governmental body with jurisdiction over at least most of the area over which costs and benefits of the policy accrue. Stormwater from one municipality can affect many other communities within its watershed, and those effects often spill over municipal, county, and state lines. Therefore, although local governments are best situated to administer land-based environmental protection requirements, the federal government has a useful role to play in ensuring that public waters are protected and market failures are overcome.

Controlling externalities makes economic sense if the net benefits of the solutions outweigh the administrative and resource costs needed to accomplish the task. This chapter evaluates the likely net benefits of enhanced stormwater-control requirements to clarify whether there is a good case for federal intervention.

As noted above, existing federal regulations of stormwater runoff focus on the construction phase of development projects. Here, we analyze specific proposals that would extend current federal stormwater management policies for development beyond the existing Phase I and Phase II stormwater management requirements. Those extensions would require postdevelopment hydrology to approximate predevelopment hydrology to neutralize the effects on stream flows and riparian ecosystems; they would also apply stringent numeric quality standards to the runoff from construction sites. We address the question of whether low impact development measures can effectively address construction-phase needs while providing long-term stormwater benefits that conventional construction-phase barriers and filters fail to provide.

We begin with a review of the USEPA's recent economic analyses of proposed stormwater runoff technical guidelines for construction and development (USEPA, 2008b,c). We then develop improved and updated estimates of the costs, benefits, and likely housing-market effects of enhanced stormwater regulations that would encourage the application of LID techniques that would serve to reduce runoff and related pollutants both during construction and afterwards. Our estimation methodology relies on an interpretive assessment of relevant studies from the academic and professional literature. Finally, we discuss sources of industry resistance to LID development techniques, and identify actions that could be taken by governments and non-governmental organizations (NGOs) to increase industry acceptance of LID stormwater control and the enhanced stormwater control regulations that might demand adoption of LID.

USEPA's 2009 Proposal

Following the year 2000 court directive to develop effluent guidelines for the construction and development industry, the EPA issued four rounds of regulatory impact analyses (RIA; USEPA, 2002, 2004a,b, 2008a–c, 2009a–d). The 2004 analysis improved on the 2002 study methodologically, but it omitted consideration of postconstruction controls that had been included in the initial report. The decision to exclude postconstruction stormwater pollution control was maintained in the final regulatory proposal issued in 2009 (USEPA, 2009a–d).

The regulatory impact analyses estimated the costs and benefits of various options for stormwater control. The final rule consists of technology requirements and numeric performance standards for the turbidity of runoff from a site during the C&D period. The technology requirements apply at all construction sites and aim to control both the mobilization and discharge of sediment and dissolved pollutants. They call for measures that will:

- Minimize the area of land disturbed.
- Implement erosion controls on disturbed areas.
- Minimize soil compaction.
- Maintain natural buffers around surface waters.
- Direct runoff to vegetated areas to maximize infiltration and sediment capture.

- Stabilize disturbed soils immediately on completion of grading activities.
- Minimize discharges of chemical spills and leaks and wash water.

The numeric standards are more limited in their application than the technology standards. The numeric standards apply only to construction sites where at least 10 acres are disturbed.[*] They require the turbidity of runoff water produced by moderate-intensity storms[†] not to exceed an average of 280 nephelometric turbidity units (NTU). In addition, developers are required to monitor the turbidity of stormwater discharges from these sites.

For economic evaluation, the measures considered by the EPA to reduce sedimentation in waterways were considered add-ons to conventional C&D practices. The cost of the regulation consists of the incremental costs of these measures. The EPA's (2009c, p. 203) estimated cost for the final regulation is $480/ton of sediment reduction (2008 dollars) and the estimated benefit is $185/ton plus other benefits that could not be monetized. The gross annual costs and benefits are estimated respectively to be $958.7 million and $368.9 million plus nonmonetizable benefits. The regulation is expected to reduce sedimentation by slightly less than 2 million tons/year.

The EPA gave little consideration to the possibility of modifying the underlying construction and development strategies to achieve the desired results instead of using add-on controls. Such modifications appear likely to reduce the calculated costs, as is shown later in this chapter.

In addition to focusing on add-ons to the development process rather than strategic changes in how development occurs, the EPA considered only the costs and benefits of measures undertaken during the active C&D phase. The potential for long-term water resource benefits beyond the C&D period were largely excluded from consideration. The exception to this statement surrounds the EPA's ultimate rejection of "active treatment systems" (ATSs) as the recommended compliance measure for large developments. ATSs use chemical flocculants to accelerate settling rates in sediment basins. In justifying its decision not to base its numeric standards on the use of ATS technologies, the EPA noted that such an approach might "present a disincentive for site planners to select controls that may be more effective from a hydrologic standpoint to maintain the predevelopment hydrology of the site" (USEPA, 2009c, p. 123). In essence,

---

[*] During a phase-in period, the numeric requirement applies to disturbed areas of 20 acres or more. The 10-acre size provision will become effective in 2014.

[†] The numeric limits are applicable to runoff from storms up to the 24-hour rainfall amount that occurs in any given location with 50% annual likelihood (i.e., a 2-year, 24-hour storm).

the EPA reasoned that, if large sediment basins were created in response to a requirement to use ATSs, it would be cost-prohibitive to remove them after the C&D phase in favor of distributed, infiltration-oriented measures that better replicate the hydrological performance of undisturbed land. Thus, in at least this respect, the EPA's choice of C&D-phase regulations considered not only the benefits realized during the period of regulation, but also the possible longer-term benefits. In so doing, the EPA affirmed the idea that strategic measures designed to minimize development impacts and retain or enhance distributed infiltration could contribute to both short- and long-term water quality improvements.

However, the EPA's 2009 rule did not impose limits on runoff nor hydrological disruption after construction is complete. That choice may be a legacy of cost–benefit analysis of postconstruction controls in earlier RIAs. For example, the EPA's (2008b) cost–benefit analysis of enhanced stormwater control regulations did not include benefits that would accrue in perpetuity from the permanent mitigation of stormwater flows that might be accomplished during the C&D process, assumed stormwater-control improvements would be done with controls added on to existing technology rather than by use of more cost-effective fundamental stormwater-control redesigns, and did not include estimates of several potentially important sources of benefits from stormwater control such as the amenity value to property owners of clarity improvements in nearby surface waters. On net, these decisions led to overstated costs and understated benefits of enhanced postconstruction-phase stormwater control regulations. A report by the National Research Council (Committee on Reducing Stormwater Discharge, 2008) recommended that the EPA develop regulations to control stormwater flows from new and redeveloped impervious surfaces using techniques that promote rainwater harvest, infiltration, and evapotranspiration. In its final rulemaking, the EPA stated a commitment to pursuing that recommendation with a final rule in place by the end of 2012 (USEPA, 2009c, p. 47).

Regulation of C&D-phase stormwater should induce some adoption of improved site design-based practices in lieu of, or in combination with, runoff capture measures. Adoption of the measures that deliver longer-term benefits would probably increase if the regulations applied to postconstruction runoff as well. The benefits associated with these outcomes would rightly be counted as benefits of the strategies induced by the regulations. In addition, as we show below, the costs of distributed and infiltration-oriented measures may actually be less than the cost of the standard postconstruction stormwater management measures. In short, regulations that encourage the use of low-impact measures for construction and development have the potential to reduce the overall costs of new development and ongoing stormwater utilities while increasing water resource benefits.

LID and Stormwater Management

LID stormwater management techniques have the potential both to reduce the costs and increase the benefits of stormwater-control activities relative to the estimates put forth in the EPA's most recent analyses. Here we evaluate those costs and benefits based on a review of the growing literature that studies LID development techniques.

Costs of LID Stormwater Control

This section uses an applied review of the literature on the costs and results of LID development techniques to evaluate the compliance costs of on-site stormwater management in new development with emphasis on the potential for LID practices to reduce those costs. LID measures for stormwater management include development plans that minimize site disturbance and preserve pervious open spaces as well as the use of filter strips, swales, infiltration trenches, green roofs, rain gardens, permeable pavements, narrower streets, downspouts connected to cisterns or rain barrels for on-site storage and reuse, and public education programs supporting the adoption and effective use of these measures.[*] The emphasis is on decentralized, on-site management measures that promote infiltration rather than relying almost exclusively on rapid conveyance to receiving waters. LID technologies are of particular interest because, in contrast to conveyance-based stormwater management systems, they can mitigate quality impairments as well as diminish the quantity of runoff (CWP, 2007b). Furthermore, unlike temporary measures to contain stormwater and trap sediments during the construction phase, LID measures can permanently curtail the stormwater flows (and related environmental impacts and public service costs) associated with newly developed areas.

Of course, during construction, added measures might be required. However, such measures are already required by current stormwater regulations and so would not represent an additional cost from enhanced stormwater regulations. Our evaluation assumes that temporary controls combined with LID design measures will enable developers, builders, and property owners to achieve the proposed pollution and runoff controls during both the construction and postconstruction phases of new development.

---

[*] See generally: Committee on Reducing Stormwater Discharge (2008), Low Impact Development Center (2007), MacMullan and Reich (2007), Narayanan and Pitt (2006), NAHB Research Center (2003), and USEPA (2007). The term LID is sometimes used more expansively to include energy- and material-conserving building design and transport-reducing site layout, as well as site-based stormwater management. The energy, material, and transport aspects of LID are beyond the scope of this chapter.

A growing body of evidence indicates that LID measures used in lieu of or integrated with conventional stormwater infrastructure can actually reduce the overall costs of stormwater management while producing quality and quantity benefits following, as well as during, the C&D period (ECONorthwest, 2007; USEPA, 2005, 2007; USEPA and Low Impact Development Center, 2007).* The conventional approaches are either conveyance-oriented curb–gutter–pipe systems or detention-oriented "best management practices" (BMPs). Drawing on this contemporary understanding of LID, the focus here is on comparisons between LID practices and conventional stormwater management rather than the temporary add-ons, such as sediment curtains or straw-bale filters, required during construction by both approaches. According to standard principles of benefit-cost analysis (e.g., Gramlich, 1990), if LID measures are adopted in response to enhanced stormwater guidelines and they reduce the overall costs of stormwater management, then the overall savings can be attributed to the policy enhancement.

Table 3.1 summarizes studies of stormwater management costs.† These studies are classified according to whether the project was new or a retrofit and whether the analysis was based on engineering design templates or case studies of actual projects. Each study is characterized by a summary estimate of the percentage difference between the costs of LID and those of the alternative.

Most of the studies summarized in Table 3.1 conclude that LID measures, other than green roofs, actually cost less than conventional stormwater management and detention-based BMPs. The cost savings are realized largely by reducing or avoiding the need for subsurface pipes and related infrastructure. Additional savings are realized when portions of sites are left in a natural state, thereby avoiding grading and planting costs; buildings are clustered so as to concentrate the site preparation work in a smaller area and simplify the road network; streets and sidewalks are narrower and require less concrete; roadside swales are used in place of curbs and gutters; land that would otherwise be reserved for centralized stormwater detention becomes available for additional lots; and LID measures diminish storage capacity needs for combined sewer systems. In a recent review, the EPA (2007) found that LID measures cost less than conventional stormwater management technologies in 11 of the 12 cases amenable to cost comparisons.

---

* Indeed, the EPA recognized this as early as its February 2002 analysis: "Many in the construction industry have found they face lower development costs with LID than with conventional `curb and gutter' design." (USEPA, 2002, pp. 2–47)

† This review is limited to studies that attempt to compare LID to conventional stormwater controls. Although it is necessarily selective, we believe the studies chosen represent current understandings of cost and performance within the stormwater engineering community.

TABLE 3.1

Review of Comparative Stormwater-Control Cost Studies[a,b]

Source	Type of Development	New or Retrofit	Study Type[c]	% Cost Difference
CRI (2005)	Commercial sites, except green roofs	New	DT	+2.3
CRI (2005)	Commercial sites, with green roofs	New	DT	+25
CRI (2005)	Seattle streetscape retrofits	Retro	CS	−25
CRI (2005)	NE IL 30–40 acre subdivisions	New	CS	−24
CRI (2005)	SW WI subdivision construction	New	DT	−27
CWP (1998)[f]	VA Piedmont medium-density residential development	New	DT	−20
CWP (1998)[f]	MD Eastern Shore low-density residential development	New	DT	−5
CWP (1998)[f]	Commercial office park	New	DT	−17
Landphair (2001)	TX highway construction	New	DT	−75 to −90
Liptan & Brown (1996)	Portland, OR office and parking lot	New	CS	−71
Liptan & Brown (1996)	Portland, OR educational facility and parking	New	CS	−68
Liptan & Brown (1996)	Portland, OR row houses	New	CS	−44
Liptan & Brown (1996)	Portland, OR 5-acre riparian development	New	CS	−27
Narayanan & Pitt (2006)	Alabama 250-acre industrial site	New	DT	−80
Tyne (2000)	Arkansas suburban residential development	New	CS	−30
NRDC (2001)	NC army base buildings and parking lot	New	CS	−20
USEPA (2002)	Hypothetical 7.5-acre residential development[d]	New	DT	<<1[e]

[a] Except where noted, studies compare costs of LID to conventional stormwater control methods (not best management practices (BMPs)) for new (not retrofit) construction.

[b] Except where noted, cost elements include only site development, infrastructure, and landscaping.

[c] Cost estimates based on: DT = design templates with architectural/engineering estimates; CS = actual case studies.

[d] Cost elements include building construction in addition to site development, infrastructure, and landscaping; construction phase only. Compares LID costs to BMP costs.

[e] Builder-developer profits estimated to decrease up to 0.8% if unable to pass any costs through to buyers. If all costs can be passed through, prices estimate to increase up to 0.09%.

[f] Center for Watershed Protection (1998), referenced in CRI (2005).

LID design is now widely recognized to yield good stormwater control at low cost, however, it is commonly recognized these techniques require careful attention to site conditions and layout before development and construction begin. The aim is to take advantage of, and even enhance, the natural capacity of a site to retard runoff and promote infiltration. The added efforts in planning and design could offset some of the cost advantages of LID. Nevertheless, some of the studies listed in Table 3.1 (e.g., Conservation Research Institute, 2005) include the costs of additional design work and still report net overall cost savings.

The studies reviewed in Table 3.1 and by the EPA (2007) focus on initial costs. It is important also to consider long-term maintenance costs. The maintenance costs for LID measures (e.g., weeding rain gardens, clearing debris from rain barrels, and unclogging porous pavement) may well be lower than the costs of maintaining detention ponds, stormwater conveyance pipes, and other conventional technologies (Hager, 2003). However, because LID measures are often located on private property whereas conveyance systems are largely public, maintenance may be more difficult to monitor and accomplish under LID design (Montalto et al., 2007). Thus, some fraction of the cost savings achieved with LID may need to be devoted to encouraging maintenance of stormwater management systems on private property.

In summary, when considering both construction phase costs and long-term maintenance expenses, LID stormwater-control technologies have the potential to be less costly than conventional collection and conveyance infrastructure. LID measures for new construction have cost advantages even if the planning and design costs are greater for LID than for conventional construction designs. Hence, this literature-based cost analysis of LID practices suggests that the cost estimates in the USEPA (2008b) are an upper bound on the costs that would actually be incurred if LID measures were included among the compliance technologies.

Benefits of LID Stormwater Control

Estimates of the benefits of LID-based stormwater control depend on whether LID measures can advance water quality goals. United States and international organizations are compiling and documenting a growing body of research concerning the effectiveness of LID practices (Landers, 2006: Low Impact Development Center, 2007). Most of this work focuses on individual practices rather than combinations of practices (see the studies reviewed by Montalto et al., 2007). However, the growing body of evidence is supportive of the effectiveness of LID measures.

The University of New Hampshire Stormwater Center (2006), for example, tests stormwater control techniques in parallel at the same site. Their data for 2005 indicate that several LID technologies—gravel wetlands and bioretention systems—provide flow control competitive with retention ponds, and LID control methods have better contaminant removal performance than conventional stormwater control methods. The Center for Watershed Protection (CWP) reviews the literature on the effectiveness of conventional as well as LID stormwater measures (CWP, 2007); see Table 3.2. The CWP's report makes clear that it is more cost-effective to address stormwater issues during initial construction when there is flexibility in site development options than to control stormwater flows with retrofits. The EPA's recent review of the LID literature concludes that "LID practices can reduce both the volume of runoff and the pollutant loadings discharged into receiving waters. . . . Reductions in pollutant loadings to receive waters, in turn, can improve habitat . . . enhance recreational uses . . .[and] decrease stormwater and drinking water treatment costs. . . ." (USEPA, 2007, p. 7).

The environmental benefits of stormwater management stem from minimizing water quality impairments and enhancing protection of habitats and aquifers.* Collateral benefits include reduced flood damage, reduced water supply costs, and avoided costs for conventional stormwater collection, conveyance, storage, and treatment systems. In the following subsections, we offer estimates of three categories of economic benefits produced by LID measures for stormwater management in new development: (1) water quality benefits, (2) reduced flood losses and infrastructure costs, and (3) savings in costs of combined sewer overflow mitigation. Our

TABLE 3.2

Summary of Stormwater Management Practice Efficiency: Median Percentile Efficiency

Management Practice	Pollutant[a,b]							
	TSS	TP	Sol P	TN	NOx	Cu	Zn	Bacteria
Dry pond	49	20	–3	24	9	29	29	88
Wet pond	80	52	64	31	45	57	64	70
Wetland	72	48	25	24	67	47	42	78
Filtering	86	59	3	32	–14	37	87	37
Bioretention	59	5	–9	46	43	81	79	N/A
Infiltration[c]	89	65	85	42	0	86	66	N/A
Open-channel[d]	81	24	–38	56	39	65	71	–25

Source: Adapted from Center for Watershed Protection (2007, pp. 6–9).

[a] Pollutant abbreviations: TSS = total suspended particulates; TP = total phosphorus; Sol P = ortho-phosphorus and dissolved phosphorus; TN = total nitrogen; NOx = Nitrogen as nitrate or nitrite; Cu = Copper; Zn = Zinc.

[b] Negative numbers indicate increases in pollutant loads.

[c] Infiltration includes infiltration trenches and porous pavement.

[d] Open-channel refers to grass channels and wet and dry swales.

---

* For an overview of ecosystem services associated with water, see Brauman et al. (2007).

approach reflects an important distinction between construction-phase measures and LID strategies: the benefits of the former occur only during the period of development whereas the benefits of the latter can endure for many years.

Willingness to Pay for Water Quality Improvements

We first evaluate the national water quality benefits of enhanced stormwater management regulation. In the 1990s, the EPA supported research using contemporary methods and data to estimate water quality benefits (Magat et al., 2000). That study carefully applied survey-based choice analysis methods to elicit tradeoffs in monetary terms between different water quality conditions and annual living costs. Based on survey samples drawn in North Carolina and Colorado, it concluded that households in those states would pay an annual average of $22.40 for a 1% improvement in generic water quality. Viscusi, Huber, and Bell (2004) expanded the survey to a national sample, and found nearly identical results: national mean annual household willingness to pay $23.17. The latter study showed the findings are robust to various econometric specification issues. Using the consumer price index (U.S. Department of Labor, 2009) for adjustment, the mean estimate of Viscusi et al. would be $30.70/year (in 2008 dollars) for a 1% change in water quality.

We assume that LID measures would perform at least as well as the construction phase detention measures in protecting against water quality degradation. The EPA (2009c, p. 194) determined that the construction-phase measures would achieve a 0.7% average reduction in suspended sediment concentrations. Adjusting proportionally the willingness to pay estimate derived from Viscusi et al. produces an estimate of $21.50/year per household.[*]

The estimates of Viscusi et al. are for average annual benefits for all households in the United States, irrespective of location. In the year 2000, there were 105.5 million U.S. households. Using these figures, the aggregate annual willingness to pay for measures that would improve water quality by 0.7% everywhere 20 years is $2.3 billion/yr. However, the EPA estimated that reduced development impacts would improve water quality in only 12% of U.S. river segments each year. Of course, the fact that development is occurring means that these river segments are probably surrounded disproportionately by households. Furthermore, the hedonic property value literature indicates that people who live nearer water value

---

[*] Magat et al. extrapolated linearly for a water quality change of more than 1%; the potential for error is less for an extrapolation to a change of less than 1%. The EPA (2009c, pp. 194–195) estimates that suspended sediments would decrease from an average of 289 mg/l without the proposed regulation to 287 mg/l with the regulation, approximately 0.7%. The averages mask substantial variation from watershed to watershed.

its quality more than average (e.g., Walsh, 2009, and Walsh, Milon, and Scrogin in Chapter 6 of this book). If we assume that 25% of U.S. households live near the affected surface waters, and that their average value of improvements is 10% above the national average, the resulting aggregate annual value of improvements is nearly $624 million.

The benefit estimate would increase further if we allowed for the fact that LID measures have enduring rather than temporary effects on water quality, unlike the construction-phase retrofit technologies. Thus, with LID, a fraction of the annual benefits would recur year after year. An ongoing nationwide benefit of only $23 million/year (less than 4% of the initial benefit) for 20 years, discounted at a real interest rate of 3%, would suffice to close the gap between our estimated willingness to pay (WTP) of $624 million/year and the EPA's estimated cost of $959 million/year (EPA, 2009c).

Flood Reduction and Infrastructure Benefits

The water quality benefits are only the beginning of the story for LID measures to manage stormwater. Next, we evaluate flood reduction and infrastructure benefits that might result from stormwater regulation.

In addition to sedimentation and water quality, in their review of the downstream benefits of stormwater management, Braden and Johnston (2004) identify benefits from reductions in downstream flooding, savings in downstream infrastructure costs, increased aquifer recharge, additions to sustainable water supplies, and stream habitat stabilization due to steadier base flows. They estimate the flood benefits alone to be in the range of 2 to 5% of the value of properties within the floodplain. These benefits would result from increases in property values when homes are less exposed to flooding or no longer need to buy flood insurance. They identify significant potential for savings in infrastructure costs. Additional benefits may accrue with added aquifer recharge and habitat improvements. Braden and Johnston decline to assign dollar values to these potential outcomes because variation from place to place in aquifer recharge rates and habitat conditions complicate the estimation, and the discounting that would accompany time lags between infiltration, aquifer response, increased water supply, and habitat effects diminish their economic significance.

Johnston, Braden, and Price (2006) apply the framework outlined by Braden and Johnston (2004) in a case study of a new regional plan for a rapidly growing area west of Chicago. They describe conventional and LID development templates and simulate storm flows in the stream network of the area. They map differences in flooded areas and compute changes in property values for the homes that would face reduced flooding in the LID scenario. They compute the sizes and costs for culverts that would be required to convey the respective storm flows. The savings are then attributed to each developed acre. Results are summarized in Table 3.3. The estimated flood benefits were $110 to $158 per developed acre and infrastructure cost savings were $340 per developed acre (all numbers are year 2000 dollars). Across the two categories, the total estimated benefits were $450 to $498 per developed acre. Adjusted to 2008 dollars, the total benefits are $563–$623 per developed acre.

The estimates of Johnston et al. are capitalized values. To convert these to annual equivalents, assume a 20-year annualization period and a 3% inflation-free interest rate. The resulting annualized benefits are approximately $40/developed acre/year. EPA assumed an average of 850,000 acres developed each year. The nationwide flood reduction and infrastructure downsizing benefits of LID measures would be on the order of $34 million/year.

CSO Savings

The benefits described above would apply to any type of development. A third category of benefit is available primarily in cities with combined sewers. This is the potential for LID measures to diminish the need for costly storage and treatment of combined sewer overflows (CSOs). Explicit treatment of CSOs is given in other chapters in this book (see, e.g., Goddard, in Chapter 10). These overflows are regulated by the USEPA, and many older U.S. cities face billions of dollars in additional expenditures unless the flow rates can be reduced. Thurston et al. (2003) find that LID measures would cost less than half as much as additional CSO storage capacity in a basin in Cincinnati. For the City of New York, Plumb and Seggos (2007) find an even greater differential in the cost-effectiveness of LID street designs, rain barrels, and increased tree planting. In a modeling study of an area in Brooklyn, New York, Montalto et al. (2007) show under various conditions that LID measures combined with a basic level of storage can be cheaper than building larger storage facilities. The predicted savings depend on the desired level of control. Considered together, these analyses suggest that integrating LID into CSO management can reduce system costs. However, the precise cost savings depend on the incremental costs of additional CSO storage capacity.

Other Benefits

Yet other types of water quality benefits from enhanced on-site stormwater management are possible. They include improvements in aquatic ecosystems made possible by reduced water temperatures and improved stream flow dynamics, resulting improvements in fishing and water-based recreation, and reduction in urban ambient temperatures and

TABLE 3.3

Review of Studies Estimating Economic Benefits of Stormwater Management

Source	Study Impetus	Type of Benefit	Underlying Methodology	Est. Benefit [year $ where applicable]
Ribaudo (1986)[a]	Value of soil conservation programs	10 types of sediment impacts[b]	Various (BT)	Avg. PV = $2.41/ton sediment lost [2000$][c]
Carson and Mitchell (1993)	Benefits of Clean Water Act	Changes in "ladder" index of water quality	Survey-based contingent valuation (Econometric)	Mean household WTP/yr: unusable → boatable: $106 boatable → fishable: $80 fishable → swimmable: $89 [1983$]
Magat et al.(2000)[e]	Value of improvements to inland water quality	Changes in % of regional (NC & CO) water bodies that are boatable, fishable, & swimmable	Survey-based choice analysis and referendum contingent valuation (Econometric)	WTP per 1% increase in proportion of water bodies meeting standards: Choice survey mean = $22.40; median = $11.30–$13.60. Referendum mean = $20.50–$27
Miles and Bondelid (2004)	Costs and benefits of construction and development regs	Changes in water quality index values	CV (Carson & Mitchell) + WQ impact modeling (BT)	WTP for avg. increase of 1.2 index points in 1.6% of river segments. Mean per household = $7.70 [2002$] Total = $13.2M[f]
Braden and Johnston (2004)[d]	Offsite value of stormwater management	Downstream flood damages	Hedonic property valuation and analysis of flood insurance costs (BT)	Partial mitigation: <2% property values Removal from 100-yr floodplain: 2–5% property values
Braden and Johnston (2004)[d]	Offsite value of stormwater management	Sedimentation	Various (BT)	0.2–0.4% of all property values; highest for waterside properties

continued

TABLE 3.3 (Continued)

Review of Studies Estimating Economic Benefits of Stormwater Management

Source	Study Impetus	Type of Benefit	Underlying Methodology	Est. Benefit [year $ where applicable]
Braden and Johnston (2004)[d]	Offsite value of stormwater management	Improved water quality	Hedonic property valuation (BT)	5% for undeveloped waterside properties; 10–15% for developed waterside residential, inclusive of sedimentation benefits
Johnston et al. (2006)	Compare off-site benefits of LID to conventional	Downstream flooding and costs of offsite drainage infrastructure	Hedonic property valuation, flood insurance rates, etc. (BT & DT)	PV = $110–$158 per developed acre [2000$] PV = $340 per developed acre [2000$]
Hansen and Hellerstein (2007)	Partial value of soil conservation programs	Opportunity cost of reservoir storage	Estimated dredging cost function (DT)	Avg. PV = $0.24/ton sediment lost [2000$]; range $0–$1.38/ton

[a] DT = design template/engineering-based; BT = benefits transfer; Econometric = statistical analysis of field data

[b] Includes: recreation, navigation, flooding, roadside and irrigation ditch maintenance, water treatment, municipal and industrial water use, steam power cooling. Excludes reservoir capacity. Variation in estimates is due, in part, to greater efficiency under the Conservation Reserve Program versus its less-selective predecessors.

[c] Updated to year 2000 dollars by Hansen and Hellerstein (2007).

[d] Estimates derived from interpretation of literature. See reference for underlying sources. Summary here includes only categories for which quantitative estimates were derived.

[e] Estimates can be recalibrated to account for different regions, types of water bodies, types of water quality improvements, and type of pollutants. Estimates exclude benefits associated with water treatment, aquatic habitat, or water body capacity.

[f] Estimates by authors of this report using a WTP function from Carson and Mitchell assuming mean WQI values of 20 for RF3Lite reaches (14%) rated boatable (WQI4 < 26), 50 for reaches (37%) rated fishable (25 < WQI4 < 70), and 75 for reaches (48%) rated swimmable (WQI4 > 70); 2002 mean household income of $61,067; U.S. resident households numbered approximately 107 million in year 2000 and value the water quality improvements in proportion to the fraction (1.6%) of surface waters predicted to benefit from the 2004 proposed rules; and no changes in attitudes toward the importance of clean water occurred between 1983 and 2002. Values in parentheses are taken from Miles and Bondelid (2004). A small number of degraded reaches are not accounted for in this calculation. The results are slightly below the lower estimate provided by USEPA (2004a) and just under 50% of the agencies higher estimate of benefits. Neither Miles and Bondelid (2004a) nor USEPA (2004a) provide a basis for determining whether affected segments are more or less populous than average watersheds. Khan (2005) argues that a 30% increase in the strength of attitudes toward clean water between 1983 (when the data underlying the value function were collected) and 2002 would be warranted and would increase the estimated WTP.

cooling costs due to reductions in impervious surfaces. Some of these benefits, especially those attributable to fishing and recreation, may be embedded in the survey-based willingness to pay estimates noted above. The inclusion of separate estimates of these benefits might result in double counting. Nevertheless, the benefits for which monetary estimates appear above probably understate the full economic benefits of LID stormwater measures.

In summary, the peer-reviewed scientific literature supports the monetization of three categories of benefits of enhanced stormwater management in newly developed areas: water quality improvement, flood damage mitigation, and stormwater conveyance infrastructure savings. Water quality benefits comprise the lion's share of the overall benefits. Assuming that LID measures perform at least as well as the construction-phase technologies in preventing pollution during active development, and that the LID measures also have enduring effects, the water quality benefits alone are of the same order of magnitude as the EPA's estimated costs ($959 million/year) of construction-phase mitigation. Embedded to some degree in these estimates are the benefits of improved aesthetics, fishing, and other recreational uses of higher-quality water. Infrastructure cost and flood damage savings associated with using LID methods to maintain or replicate predevelopment hydrology would be worth an additional $34 million/year. Total benefit estimates would be even larger if estimates of other direct benefits (such as improved wildlife populations and aquatic habitat) and ancillary benefits (such as reductions in urban heat island effects and CSO management savings) were possible.

Market Effects of LID Stormwater Control

Regulation to require enhanced stormwater controls on new construction might have direct effects on the prices and transacted quantities of houses on the market. If enhanced controls are added onto conventional stormwater controls, the cost of new home construction might increase and squeeze some low-income households out of the housing markets. For the proposed regulation, the EPA (2009b) found that less than 1% of firms would experience negative financial implications in the long run, and the employment, housing price, and affordability implications are minuscule. This finding is reinforced in the relevant literature (e.g., Dunn, Quigley, and Rosenthal, 2005; Sutton, 2002; Brasington and Haurin, 2006). However, the story may be even more benign for the adoption of LID measures. Based on the review of cost studies presented in this chapter, the costs of permanent on-site stormwater management measures may well be negative; that is, LID practices may actually increase profits. If so, then these measures could actually reduce prices and expand access to housing. The simple graphical analysis given in panel (a) of Figure 3.1 provides

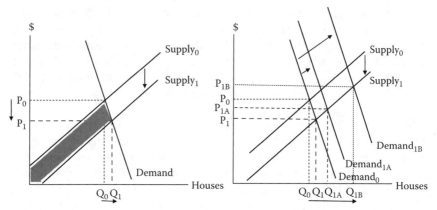

Panel (a): Decreased Construction Costs Only   Panel (b): Decreased Construction Costs,
                                                              Increased Demand

FIGURE 3.1
Market effect of decreased construction costs.
Notes: 1) The demand curve gives the number of houses that consumers will purchase at a given price. 2) The supply curve gives the number of houses that developers will offer for sale at a given price. 3) Decreased construction costs shift the supply curve down; in panel (a), market price falls from P_0 to P_1 and output increases from Q_0 to Q_1. 4) Total social welfare in panel (a) increases by the shaded area. 5) If LID stormwater control yields ancillary benefits to homeowners, market demand shifts up. We show two scenarios: $Demand_{1A}$ and $Demand_{1B}$. 6) The demand shift magnifies the increase in market output; both Q_{1A} and Q_{1B} are larger than Q_1. 7) Small demand shift yields market price $P_1 < P_{1A} < P_0$. Large demand shift yields price increases relative to the no-LID baseline; $P_{1B} > P_0$.

a qualitative understanding of the market effects of such a change. If LID stormwater-control methods reduce construction costs, then the supply curve for new houses shifts down: more houses will be built by developers for a given market price. The new market equilibrium will feature lower prices and more houses built. Total social welfare increases as represented by the shaded area. Consumers are unambiguously better off, and developers may be better off as well if the cost savings and increased output dominate the fall in market prices.

In addition to lowering stormwater-control costs, LID design may produce localized ancillary benefits that affect market outcomes. Consumers have been shown to be willing to pay a premium for homes that benefit from neighborhood water features such as detention basins or wetlands that are part of stormwater management programs (Earnhart, 2001; USEPA, 2005). Additionally, some studies have found consumers willing to pay more for homes in clustered developments than for homes in conventionally designed subdivisions, and consumers perceive value in the additional landscaping that is an integral part of LID design (USEPA, 2007).

On-site stormwater control measures may increase maintenance costs borne directly by homeowners. However, on balance the evidence indicates that consumers' may likely be willing to pay more for homes served by these technologies. Thus, the market analysis shown in panel (a) of Figure 3.1 is incomplete, inasmuch as it considers only the direct effects of changes in construction costs. As shown in panel (b) of Figure 3.1, these postconstruction effects will cause the demand curve for housing built under the proposed stormwater regulations to shift up; that is, the quantity of new housing demanded at a given price goes up. The net impact of reduced construction costs and reduced homeownership costs on housing prices is indeterminate; market prices may fall or rise depending on whether the supply shift or the demand shift is dominant. In any scenario, the demand-side effects serve to increase unit sales and increase both consumer and producer welfare over the scenario in which only changes in construction costs were considered.

This market analysis implies that significant negative employment effects are unlikely to result from regulations that encourage developers to incorporate LID stormwater management measures in new development projects. In fact, such regulations could actually yield increased employment in the development industry. If construction costs associated with this regulation are, in fact, negative, and if postconstruction benefits to homeowners cause demand for new homes to increase, then output in the housing industry will likely go up, increasing labor demand in the industry.

Barriers to Industry Acceptance of LID, Possible Solutions, and Calls for Research

The analysis above indicates that LID stormwater controls yield large environmental benefits while reducing the costs of building homes. If this is the case, then why does the home building industry fight so hard to prevent enhanced stormwater regulations that would essentially just require adoption of LID techniques? Here we discuss some factors that may serve as hidden costs or otherwise create barriers to adoption of LID, and suggest remedies for the problems discussed.

First, many cities have zoning ordinances and building codes that create significant barriers to LID design. LID approaches often entail narrower streets and clustered houses that can violate minimum street widths and caps on housing density. Extant building codes often require impervious driving surfaces and conventional stormwater controls that are the antithesis of infiltration-based LID water management. Developers may have to file for variances or other regulatory approvals to use novel LID practices (Luger and Temkin, 2000), and approval delays compound financial risks. If developers cannot use LID techniques because of outdated building/zoning regulations, then the EPA's large cost and low benefit estimates are likely to be reasonable estimates of the consequences of enhanced storm-water control regulations.

Fortunately, this is not a difficult problem to solve. Local governments can revise their codes to make LID feasible without special approval. States, regional authorities, and NGOs can help by drafting model codes and regulations that preserve the protections communities want while making it possible for builders to use LID without having to file for end-less variances. For example, the Boston, Massachusetts Metropolitan Area Planning Council (2003) provides a checklist of "simple modifications to local codes" that "can encourage builders and property owners to apply low impact techniques, while ensuring high quality development, ade-quate access, and public safety."

A second reality of LID-based stormwater control is that it shifts both initial and ongoing responsibility for stormwater management from the city to individual builders and property owners. With current regulations and designs, builders just have to design the site to convey stormwater to the streets or storm sewer system; the city takes care of the rest. LID often features some water management infrastructure in municipal areas (e.g., green streets); however, there are green roofs, cisterns, vegetated swales, and so forth on private property that must be installed properly by the builders and maintained by the homeowners. Builders and households may be reluctant to accept this shift in responsibility when the status quo is comfortable for them.

Such reluctance could be overcome by providing developers and home-owners with incentives to stimulate LID adoption. Such incentives are the subject of other chapters in this book (see, e.g., Hodge and Cutter in Chapter 9 and Parikh et al. in Chapter 8). There is precedent in laws like the Energy Policy Act of 2005 for the federal government to offer tax cred-its to homeowners who invest in "green" upgrades to their homes; federal or state policies could use similar tax credits to encourage builders to use LID design in their developments. In towns that have stormwater-related impact fees, LID developments could be granted lower impact fees to reflect the reduced stormwater management burdens those properties place on the city. Differential impact fees would provide strong incentives for devel-opers to use LID, while not requiring budget outlays from municipalities.

A third barrier to LID adoption may simply be the new expertise required to switch from conventional to LID building design. Adopting LID requires builders to learn a significant body of new knowledge: how to use natural topography to enhance infiltration (instead of just flattening the site), what plants work best in vegetated swales for a given region, how to install and use cisterns and rain gardens as part of a site's design, and so forth. Construction and development is a highly decentralized indus-try that does not have professional requirements for (and thus low-cost access to) continuing education in new building techniques. Although

LID construction may be cheaper than conventional designs, there are major transaction costs associated with gaining the expertise necessary to utilize LID approaches.

To accelerate the diffusion of LID knowledge into the industry, states or regional authorities could develop more free training programs for builders to develop LID expertise. Governments could retain local LID consultants who could work with builders in the planning stages of new projects and advise them if problems arise during the implementation process. Such programs would be best managed at something lower than the federal level, inasmuch as effective LID design will vary with the topography, climate, soils, and horticultural potential of a given region.

Conclusions

There is a growing consensus that stormwater can be managed more effectively through on-site infiltration-oriented measures than exclusively through conventional strategies emphasizing collection, conveyance, detention, and, where needed, storage and treatment. These measures can reduce water quality impairments while mimicking predevelopment hydrology. Furthermore, when the infiltration-oriented measures are included at the outset in designing and implementing development plans, they can reduce development costs, thereby putting downward pressure on prices and increasing housing availability and access. Thus, for new development, LID technologies offer the potential to reduce runoff and associated pollution during C&D activities, as well as over the longer term, at little if any cost. The adoption of uniform national guidelines would hasten the adoption of these measures and further drive down their costs.

Not only can stormwater be managed more effectively and cheaply through LID strategies, such strategies can produce significant economic value through better water quality, reduced downstream flooding, and lesser costs of public infrastructure. Using EPA-sponsored methodology, we estimate the annual improvements in water quality alone are worth more than $600 million annually. An additional $26 million/year would be realized through reduced downstream flooding and infrastructure savings. The LID measures are durable, so some of the benefits will extend over many years. If the recurring benefits were on the order of only about $20 million/year, less than 5% of the initial benefits, then the overall benefits of the LID measures would be comparable in magnitude to the EPA's estimated costs of construction-phase palliatives. Some of the benefits would accrue to homeowners through higher market value of LID homes and to developers through greater profits. Other types of benefits for which monetization is not readily available would accompany runoff reduction from new C&D sites. However, in the presence of low or negative costs, the monetized benefits alone more than suffice to justify enhanced stormwater controls.

We conclude there is compelling evidence in support of stringent new stormwater control regulations, such as those on which the EPA is currently collecting information, that would induce widespread use of LID design in the U.S. home building industry, but our cost–benefit analysis could apply to a broader range of programs. Social well-being could be increased by nonregulatory efforts to stimulate adoption of LID design. NGOs and federal agencies could institute programs to help municipalities alter building codes to ensure that no legal barriers prevent adoption of LID development design. Such organizations could develop education programs to make it easier for home-building contractors to learn how to implement LID designs in their areas. Activities that increase adoption of LID will yield large social net benefits from improved stormwater management.

Acknowledgments

This chapter is based in part upon work supported by the project nos. ILLU-470-316 and MRF 470311 of the Illinois Agricultural Experiment Station and Cooperative State Research, Education, and Extension Service, U.S. Department of Agriculture, and by the Natural Resources Defense Council (NRDC), and the Waterkeeper Alliance (WA). Nancy Stoner of NRDC and Jeff Odefey of WA encouraged our interest in stormwater regulatory proceedings. Jesse Pritts and Hale Thurston of the USEPA provided helpful insights into the Agency's analysis of the regulation. However, any opinions, findings, and conclusions or recommendations expressed in this publication are those of the authors and do not necessarily reflect the views of the sponsors or the noted individuals.

References

Boston MAPC. (2003). *Low Impact Development: Do Your Local Codes Allow It?* (http://www.mapc.org/regional_planning/LID/LID_codes.html, accessed 1/16/2008.)

Braden, J.B., and D.M. Johnston. (2004). Downstream economic benefits from storm-water management. *J. Water Resources Plan. Manage.*, 130(6): 498–505.

Braden, J.B., and S. Proost (Eds.). (1997). *The Economic Theory of Environmental Policy in a Federal System.* Cheltenham, UK: Edward Elgar.

Brasington, D., and D. R. Haurin. (2006). Educational outcomes and house values: A test of the value added approach. *J. Regional Sci.,* 46(2): 245–268.

Brauman, K.A., G.C. Daily, T. Ka'eo Duarte, and H.A. Mooney. (2007). The nature and value of ecosystem services: An overview highlighting hydrologic services., *Annu. Rev. Environ. Resources,* 32: 67–98.

Carson, R., and R. Mitchell. (1993). The value of clean water: The public's willingness-to-pay for boatable, fishable, and swimmable quality water. *Water Resources Res.,* 29(7): 2445-2454.

Center for Watershed Protection (CWP). (1998). *Better Site Design: A Handbook for Changing Development Rules in Your Community.* Ellicott City, MD.

Center for Watershed Protection (CWP). (2007). *National Pollutant Removal Performance Database, Ver. 3.* Ellicott City, MD.

Committee on Reducing Stormwater Discharge Contributions to Water Pollution, National Research Council. (2008). *Urban Stormwater Management in the United States.* Washington, DC: National Academies Press.

Conservation Research Institute (CRI). (2005). *Changing Cost Perceptions: An Analysis of Conservation Development.* Elmhurst, IL, February.

Crowe, D. (2002). Comments in Response to EPA's "Estimation of Capital Costs for Technology Options (DRAFT Revised July 20, 2001)." Unpublished document, Washington, DC: National Association of Home Builders.

Dunn, S., J.M. Quigley, and L.A. Rosenthal. (2005). The effects of prevailing wage requirements on the cost of low-income housing. *Indust. Labor Relations Rev.,* 59(1): 141–157.

Earnhart, D. (2001). Combining revealed and stated preference methods to value environmental amenities at residential locations. *Land Econ.,* 77(1): 12–29.

ECONorthwest. (2007). *The Economics of Low-Impact Development: A Literature Review.* Portland, OR, November.

Gramlich, E.M. (1990). *A Guide to Benefit-Cost Analysis,* 2nd ed. Long Grove, IL: Waveland Press.

Hager, M.C. (2003). *Low-Impact Development: Lot-Level Approaches to Stormwater Management Are Gaining Ground.* Stormwater. http://www.forester.net/sw_0301_ low.html, accessed 10/2007.)

Hansen, L., and D. Hellerstein. (2007). The value of the reservoir services gained with soil conservation. *Land Econ.,* 83(3): 285–301.

Hayes, S.A., R.A. McLaughlin, and D.L. Osmond. (2005). Polyacrylamide use for erosion and turbidity control on construction sites. *J. Soil Water Conserv.,* 60(4): 193–199.

Johnston, D.M., J.B. Braden, and T.H. Price. (2006). Downstream economic benefits of conservation development. *J. Water Res. Plan. & Mgmt.,* 132(1): 35–43.

Khan, S. (2005). Billion Dollar Runoff: The Transfer of Benefits in the United States Government's Valuation of Stormwater Regulations for Construction and Development. MSc Paper, Imperial College, University of London.

Landers, J. (2006). Selecting Stormwater BMPs. Stormwater (May/June). (http://www.forester.net/sw_0605_selecting.html, accessed 12/2/ 2007.)

Landphair, H.C. (2001). Cost to performance analysis of selected stormwater quality best management practices. In: *Proceedings of the 2001 International Conference on Ecology and Transportation*, C.L. Irwin, P. Garrett, and K.P. McDermott (Eds.). Center for Transportation and the Environment, Raleigh, NC: North Carolina State University, pp. 331–344.

Line, D.E., and N.M. White. (2001). Efficiencies of temporary sediment traps on two North Carolina construction sites. *Trans. ASAE*, 44(5): 1207–1215.

Liptan, T., and C.K. Brown. (1996). A Cost Comparison of Conventional and Water Quality-Based Stormwater Designs. City of Portland, Bureau of Environmental Services. Portland, OR.

Low Impact Development Center. (2007). San Bernardino County PIN#9172. Low Impact Development Guidance and Training for Southern California: Literature Review (August 16 Draft). Beltsville, MD.

Luger, M.I., and K. Temkin. (2000). *Red Tape and Housing Costs*. New Brunswick, NJ: CUPR Press.

MacMullen, E., and S. Reich. (2007). *The Economics of Low-Impact Development: A Literature Review*. Portland, OR: ECONorthwest, November.

Magat, W.A., J. Huber, W.K. Viscusi, and J. Bell. (2000). An iterative choice approach to valuing clean lakes, rivers, and streams. *J. Risk Uncertainty*, 21(1): 7–43.

Miles, A., and T. Bondelid. (2004). Estimation of National Economic Benefits Using the National Water Pollution Control Assessment Model to Evaluate Regulatory Options for the Construction and Land Development Industry. Research Triangle Park, NC: Research Triangle Institute, draft report prepared for USEPA, February.

Montalto, F., C. Behr, K. Alfredo, M. Wolf, M. Arye, and M. Walsh. (2007). Rapid assessment of the cost effectiveness of low impact development for CSO controls. *Landscape Urban Plan.*, 82: 117–131.

Narayanan, A., and R. Pitt. (2006). Costs of Urban Stormwater Control Practices. Unpublished manuscript, Department of Civil, Construction, and Environmental Engineering, University of Alabama, Tuscaloosa, January.

National Association of Home Builders (NAHB) Research Center. (2003). The Practice of Low Impact Development. Upper Marlboro, MD. Report to the U.S. Department of Housing and Urban Development, July.

Natural Resources Defense Council (NRDC). (2001). Better Parking-Lot Design. Stormwater Strategies: Community Responses to Runoff Pollution. (http://www.nrdc.org/water/pollution/storm/chap7.asp#FBRAG accessed January 11, 2010.)

Oates, W. (2001). A Reconsideration of Environmental Federalism. Discussion Paper 01-54. Washington, DC: Resources for the Future. (http://www.rff.org/documents/RFF-DP-01-54.pdf, accessed 12/29/2007.)

Plumb, M., and B. Seggos. (2007). Sustainable Raindrops: Cleaning New York Harbor by Greening the Urban Landscape. Riverkeeper, Tarrytown, NY. (riverkeeper.org/special/Sustainable_Raindrops_FINAL_2007-03-15.pdf, accessed 12/ 21/2007.)

Ribaudo, M.O. (1986). Reducing Soil Erosion: Offsite Benefits. U.S. Department of Agriculture, Economic Research Service, AER 561, Washington, DC.

Sutton, G.D. (2002). Explaining changes in house prices. *BIS Quart. Rev.*, September, 46–55.

Thurston, H.W., H.C. Goddard, D. Szlag, and B. Lemberg. (2003). Controlling storm-water runoff with tradable allowances for impervious surfaces. *J. Water Res. Plan. & Mgmt.*, 129(5): 409–418.

Tyne, R. (2000). Bridging the gap: Developers can see green, economic benefits of sustainable site design and low-impact development, *Land Development* 13(1): 27–31.

University of New Hampshire Stormwater Center. (2006). 2005 Data Report. (http://www.unh.edu/erg/cstev/pubs_specs_info/annual_data_report_06.pdf, accessed 12/28/2007).

U.S. Department of Agriculture (USDA). (2007). Conservation Policy: Farmland and Grazing Land Protection Programs. Economic Research Service, Washington, DC. (www.ers.usda.gov/Briefing?Conservation Policy/farmland.htm, accessed 11/29/2007.)

U.S. Department of Labor, Bureau of Labor Statistics. (2009). CPI Detailed Report: Data for November 2009. Table 24. Historical Consumer Price Index for All Urban Consumers. (CPI-U):U.S. City Average, All Items. Washington, DC. http://www.bls.gov/cpi/cpid0911.pdf (accessed January 12, 2010.)

USEPA. (1997). Urbanization and Streams: Studies of Hydrologic Impacts. Washington, DC., EPA 841-R-97-009. (http://www.epa.gov/owow/nps/lid/lidlit.html, accessed 12/21/2007.)

USEPA. (2002). Economic Analysis of Proposed Effluent Limitation Guidelines and New Source Performance Standards for the Construction and Development Category. OMB Review Draft.

USEPA. (2004a). Economic Analysis for Final Action for Effluent Guidelines and Standards for the construction and Development Category. Washington, DC: Office of Water, EPA-821-B-04-002.

USEPA. (2004b). Effluent Limitations Guidelines and New Source Performance Standards for the Construction and Development Category. 69 Federal Register 22472-22483 (April 26).

USEPA. (2005). Low-Impact Development Pays Off. Nonpoint Source News-Notes, No. 75, Washington, DC, May. (http://www.epa.gov/NewsNotes/issue75/75issue.pdf, accessed 12/2/ 2007.)

USEPA. (2007). Reducing Stormwater Costs through Low Impact Development (LID) strategies and Practices. Washington, DC, EPA 841-F-07-006.

USEPA. (2008a). Development Document for Proposed Effluent Guidelines and Standards for the Construction and Development Category. Office of Water, Washington, DC.

USEPA. (2008b). Economic Analysis of Proposed Effluent Limitation Guidelines and Standards for the Construction and Development Industry. Office of Water, Washington, DC, EPA-821-R-08-008.

USEPA. (2008c). Environmental Impact and Benefits Assessment for Proposed Effluent Guidelines and Standards for the Construction and Development Category. Office of Water, Washington, DC, EPA-821-R-08-009.

USEPA. (2008d). Proposed Effluent Limitations Guidelines and Standards for the Construction and Development Industry. Washington, DC, Docket No. EPA-HQ-OW-2008-0465.

USEPA. (2009a). Development Document for Final Effluent Guidelines and Standards for the Construction and Development Category. Washington, DC, EPA-821-R-09-010.

USEPA. (2009b). Economic Analysis for Final Effluent Guidelines and Standards for the Construction and Development Category. Washington, DC, EPA-821-R-09-011.

USEPA. (2009c). Effluent Limitations Guidelines and Standards for the Construction and Development Point Source Category: Final Rule. Code of Federal Regulations, 40, part 450.

USEPA. (2009d). Environmental Impact and Benefits Assessment for Final Effluent Guidelines and Standards for the Construction and Development Category. Washington, DC., EPA-821-R-09-012.

USEPA and Low Impact Development Center. (2007). Low Impact Development (LID): A Literature Review. Washington, DC, EPA-841-B-00-005. (http:// www.epa.gov/owow/nps/lid/lid.pdf, accessed 12/2/2007.)

Viscusi, W.K., J. Huber, and J. Bell. (2004). The value of regional water quality improvements. Discussion Paper No. 477, John M. Olin Center for Law, Economics, and Business, Harvard Law School, July.

Walsh, P. (2009). Hedonic Property Value Modeling of Water Quality, Lake Proximity, and Spatial Dependence in Central Florida. Unpublished PhD Dissertation, University of Central Florida, Orlando.

4

Accounting for Uncertainty in Determining Green Infrastructure Cost-Effectiveness

Franco A. Montalto, Christopher T. Behr, and Ziwen Yu

CONTENTS

Introduction

As was mentioned in the last chapter and in others before it, uncertainty about the effectiveness of low-impact development practices and green infrastructure as stormwater runoff control measures is an area of research focus. In this chapter, the analysis is taken a step further by linking the uncertainty in green infrastructure (GI) effectiveness with uncertainty that is inherent in GI costs and municipal implementation programs. The chapter demonstrates that in order to make GI a central component of their infrastructure programs and land use plans, municipalities, water utilities, and other natural resource planners need to better account for such uncertainty. Because GI programs involve landscape-scale assessments and stakeholder participation, traditional infrastructure decision-making processes do not easily apply. Instead, a five-step modeling approach is outlined that integrates the uncertainties in hydrological, economic, and key stakeholder interests and perspectives to provide the policy makers with information that best enables them to evaluate tradeoffs in risk and cost.

Background

Proponents of a new generation of decentralized stormwater management practices claim the negative impacts of urbanization on terrestrial and aquatic ecosystems can be significantly reduced through improved designs of land parcels, even in densely populated urban watersheds. GI practices, known alternatively as low-impact development or best management practices, attempt to mitigate the impacts of urbanization by "mimicking" natural hydrological processes of water harvesting, infiltration, evapotranspiration, and recycling on individual developed sites. GI practices diverge conceptually and practically from conventional "mechanistic-hydrotechnical approaches" to stormwater management (Zalewski, 2000) that address pollution problems at the end-of-pipe, but fail to "solve" larger environmental problems in the catchment. The USEPA and other public and private agencies affirmed their support for GI programs in the recent report *Green Infrastructure Statement of Intent*, which discussed how GI strategies that use "soil and vegetation" are both more "cost-effective" and "environmentally preferable" to "centralized hard infrastructure" (USEPA, 2007). Water utilities are beginning to follow suit. For example, in updates to its Combined Sewer Overflow (CSO) Long Term Control Plan submitted in 2009, the Philadelphia Water Department proposed a $1.6 billion investment in GI over 20 years as a core component of its watershed-based approach to compliance with the EPA CSO Control Policy (Smart Growth Network, 2010). New York City's 2010 Green Infrastructure Plan commits to, among other things, controlling runoff from 10% of impervious surfaces in CSO drainage areas with GI, also over a period of the next 20 years (NYC, 2010).

The viability of such programs is contingent upon the ability of GI to achieve cost-effective reductions in urban runoff. Cost-effectiveness can be measured in various ways, but essentially involves comparing the effectiveness (or performance) of a GI system in capturing stormwater (over a given period of time) to the cost of retaining or capturing these volumes. Cost-effectiveness evaluations are performed for individual GI installations, but must also be scaled up to the watershed level and tracked over time for comparison with conventional "gray" approaches to stormwater management. Cost-effectiveness calculations do not typically capture the full value of stormwater capture, as some studies have shown (Philadelphia Water Department, 2009), but provide sound financial and performance comparisons among GI systems and between GI and conventional systems.

Public programs that promote GI seek through various mechanisms to encourage decentralized implementation of GI on properties across a watershed. A successful program would produce high levels of GI adoption that effectively reduce the overall rate and volume of stormwater runoff, potentially also offsetting potable water consumption where harvested rainwater is used for irrigation or other uses (Basinger, Montalto, and Lall, 2010), and enhancing the ability of the urbanized landscape to provide valuable ecosystem services. The cost-effectiveness measures apply to GI programs in which the public program cost is evaluated against the cumulative impact of individual installations across the watershed.

Many different kinds of studies (Zhou et al., 2009) and programs (USEPA, 2010) have been conducted and are underway to improve our understanding of the cost-effectiveness of GI practices. Besides helping frame the magnitude of cost-effectiveness, these studies have identified several significant areas of variability in GI systems:

- Capital and annual maintenance costs
- Hydrological performance/effectiveness
- Program performance and watershed adoption

These uncertainties can have significant implications as to how GI can be incorporated into the stormwater management programs developed by municipalities, water utilities, and watershed managers. The issues are particularly important for programs that include some cost sharing because higher levels of public-supported costs reduce the balance to be paid by private entities and lead to higher rates of adoption.

The purpose of this chapter is to discuss the nature of these three areas of uncertainty and how they influence decision making. Specifically, the chapter proposes a methodology for quantifying this uncertainty, allowing decision making to proceed with better information and with confidence that the GI program is structured and targeted most effectively. The first section of this chapter discusses the various sources of uncertainty associated with predicting the cost-effectiveness of GI and includes some practical examples. Next, the GI decision-making process is outlined, specifically with respect to uncertainty quantification, and stakeholder participation.

The chapter also addresses the need for developing new tools to aid government agencies, planners, engineers, and other design professionals in their efforts to incorporate GI into sustainable urban water management plans (Novotny and Brown, 2007). We introduce a method for estimating GI cost-effectiveness with participation of watershed stakeholders. This approach makes use of Version 2.0 of the low-impact development rapid assessment (or LIDRA) model, a web-based planning model. LIDRA 2.0 enables users to quantitatively consider the uncertainty associated with GI performance, cost, and adoption in evaluating the cost-effectiveness of GI as an urban runoff reduction measure. The end of the report discusses interpretation of uncertainty analysis for decision making.

Uncertainty in GI Cost-Effectiveness Predictions

GI cost-effectiveness is defined as the reduction in runoff achieved by a given investment in GI. An investment in GI includes the capital cost of the initial installation, as well as recurring annual costs for operation and management (O&M) over the system's useful life, together represented as a life-cycle cost (LCC). The effectiveness of individual GI systems describes by how much they reduce the quantity and rate of runoff generated from hydraulically connected geographic areas (i.e., catchments). When multiple GI systems are installed across a larger land area (e.g., a watershed), all the incremental reductions in runoff reduces the load on stormwater or combined stormwater and wastewater collection systems and can improve stream and surface water quality.

The magnitude of such impacts requires reliable information about how much GI systems cost, how well they reduce runoff, and what adoption can be expected across a particular watershed through time. Because GI is a recent technology, both cost and performance studies are limited; this is especially true when considering a specific catchment. Even fewer studies have quantified the response of watershed stakeholders to GI policies and incentive programs (Hill, 2007). The certainty with which we can predict the cost-effectiveness of a GI program is thus limited by the range of variability that is possible in GI cost, performance, and adoption. We review these three key sources of uncertainty using real-world examples in the three subsections that follow.

Variability in Costs

Cost information about GI systems come from studies and databases that have been complied national data (International Stormwater Best Management Practices [BMP] database 2010; CNT, 2010; Montalto et al., 2007), or more generic cost calculators. Table 4.1 presents high, medium, and low values of recent GI installation and O&M costs, as well as their expected durability, compiled from various national studies by the authors.

The data indicate that GI capital costs are variable, even within the same city. GI O&M costs are even more difficult to predict because in the decade or so since GI systems have begun to appear, there has been insufficient time to reliably witness, inventory, and report the required operation and maintenance tasks and the effort required to perform them.

When a GI system is being considered for a specific site, the cost can perhaps most reliably be estimated using typical construction time and materials estimates, calibrated to local market conditions and labor costs. In this way, the cost of the GI component of a project can be estimated with all other engineered elements.

At the watershed scale, the cost to build, operate, and maintain a decentralized portfolio of GI systems is more difficult to estimate because each new GI design will need to be uniquely customized to local conditions, including siting, land use, code, and other programming preferences. Costs will vary due to the method and quality of construction, which will vary from site to site. Under such variable conditions, precise cost estimates are difficult, if not impossible, to obtain.

As an example of how widely GI system costs can vary within just one city, and to enumerate some of the causes of GI cost variability, Table 4.2 contains recently completed green roof costs compiled by the authors during the development of New York City's Sustainable Stormwater Management plan. The installation costs varied from $5–$71/sf, with variability dependent on a variety of factors. These factors included whether the installation was a retrofit or associated with a new roof, which components were included in the green roof cost estimate, which green roof products were used, the depth of the growing media, the method of construction (e.g., crane versus blower), the extent of roof furniture present, and the extent to which the property owner or volunteers were engaged in the installation process.

In addition to the various sources of uncertainty in GI systems, it is also important to understand the uncertainty in conventional systems. Montalto et al. (2007) note that under current market conditions, the LCC of GI systems, as borne by a property owner/developer and eventual site owner, are typically higher than for conventional surfaces. For example, green roofs cost more than regular roofs, porous driveways may cost more than asphalt ones, and directing a downspout to a cistern is more

TABLE 4.1

Compilation of GI Costs

Name	Source	Initial Cost Low	Annual Cost Low	Lifetime Low (Years)	Initial Cost High	Annual Cost High	Lifetime High (Years)
Green Roof	Pricing Sheet from CNT	62.969($/m2)	13.455($/m2)	50	449.178($/m2)	21.528($/m2)	50
Blue Roof	National Green Values Calculator Methodology	43.055($/m2 area treated)	0($/m2 area treated)	20	43.055($/m2 area treated)	0($/m2 area treated)	20
Rain Garden	Pricing Sheet from CNT	39.504($/m2)	1.938($/m2)	100	52.743($/m2)	3.66($/m2)	100
Driveway Permeable Pavement	Pricing Sheet from CNT	17.87($/m2)	0.387($/m2)	15	107.424($/m2)	0.387($/m2)	50
Downspout to Yard	National Green Values Calculator Methodology	3.195($/m2 roof area managed)	0.006($/m2 roof area managed)	30	3.195($/m2 roof area managed)	0.006($/m2 roof area managed)	30
Rain Barrels	National Green Values Calculator Methodology	42.19($/m2 area treated)	0($/m2 area treated)	20	42.19($/m2 area treated)	0($/m2 area treated)	20
Curbside Infiltration	Pricing Sheet from CNT (Vegetated swale)	3.223($/m2)	0.646($/m2)	100	5.382($/m2)	4.628($/m2)	50
Parking Lane Permeable Pavement	Pricing Sheet from CNT	17.87($/m2)	0.387($/m2)	15	107.424($/m2)	0.387($/m2)	50
Sidewalk Permeable Pavement	Pricing Sheet from CNT	17.87($/m2)	0.387($/m2)	15	107.424($/m2)	0.387($/m2)	50
Tree	Pricing Sheet from CNT	175($/each)	20($/each)	13	400($/each)	20($/each)	37

TABLE 4.2

Installation Costs for Recently Constructed NYC Green Roofs

Project Name	Installation Cost ($)	Roof Area (sf)	$/sf
Bronx Prep Charter School	69,000	2500	28
2241 Webster Ave Fordham Bedford	47,600	2000	24
Mt. Hope Housing Co.	125,000		
W. Side Fed for Senior & Supp Housing	125,000	8000	16
St. Simon Stock Parish	5,600	501	11
St. Simon Stock School	125,000	3498	36
Abraham House 344-348 Willis Ave	48,250	2000	24
1231 Lafayette Ave; Sust. South Bronx	25,000	1501	17
Youth Ministries for Peace & Justice	49,888	700	71
WHEDCO	125,000	26,292	5
Bronx County Building	230,000	10,000	23
Bronx River Arts Center	49,740	2300	22

Source: NYC. (2008). Sustainable Stormwater Management Plan. Accessed at http://www.nyc.gov/html/planyc2030/html/stormwater/ stormwater.shtml

expensive than leaving it connected directly to the drainage system. In addition, the uncertainty in costs of conventional stormwater systems can be heavily influenced by labor, materials, and right-of-way which in the past few years have been widely varying. As such, because GI programs and individual GI implementation decisions include the cost differences between GI and conventional systems, both sources of uncertainty must be taken into account.

Variability in Performance

The performance of GI systems as stormwater controls is also variable. Some evidence of this variability is derived from monitoring studies, whereas others are based on predictions made using models or calculators. Reconciling these measures is challenging because some GI performance studies focus on water quality improvement (such as Chapter 2 in this volume), and others focus on water quantity (i.e., the ability of GI systems to detain or retain stormwater as is one of the foci of the next chapter).

The performance metric perhaps most relevant for urban areas is volume reduction. That is, by how much can GI systems reduce the volume of stormwater entering collection systems and receiving water bodies during wet weather events? This performance metric is dependent on precipitation, soil, topography, and other local conditions, as well as the unique

FIGURE 4.1
Curbside bioswales along 6th Street in the Gowanus Canal Watershed (Brooklyn, NY). (Image credit: A. Bayley.)

engineered relationship between the system's catchment area and the volume of storage available at the onset of rainfall.

To demonstrate how variable GI system performance can be, we estimate the volume of stormwater that can be captured in seven curbside bioswales to be installed in the Gowanus Canal watershed (Brooklyn, New York) as part of a recently funded GI pilot project (Figures 4.1 and 4.2).

The design is as follows: during wet weather, stormwater flowing along the curbs (indicated by the gray arrows in Figure 4.1) is directed into each of seven vegetated bioswales within the existing sidewalks. Because the elevation of the soil surface inside the bioswales is depressed below the inlet invert, stormwater that enters each system will rapidly spread over the soil surface, maximizing infiltration and evaporation capacity. If the water level in the bioswale reaches the elevation of the inlet invert, stormwater will no longer be able to enter the facility, and will instead continue flowing along the curb, bypassing the bioswale and entering the existing street-end catchbasin as it does under existing conditions. Because the street is sloped, at the downgradient end of the bioswale, a French drain and vertical riser pipe connected to an outlet will allow for the slow release of any ponded water that has not been able to infiltrate or evaporate, ensuring that no water will remain ponded on top of the soil surface for more than 24 hours (a threshold driven by vegetation health and vector-breeding considerations).

The size, number, and placement of the bioswales at this particular site are limited by the presence of driveways and underground utilities, siting constraints associated with several local relevant codes and regulations, as well as surface topography. After all of these engineering considerations are taken into account, Table 4.3 shows the dimensions of the seven

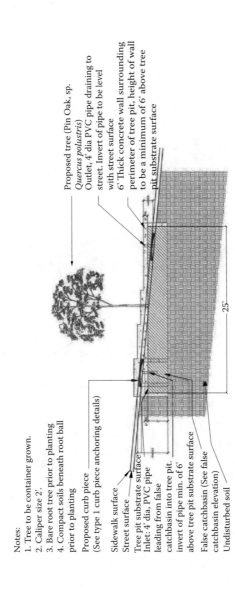

FIGURE 4.2
Curbside bioswale section (parallel to street). (Image credit: A. Bayley.)

TABLE 4.3

Performance of the Curbside Bioswales in NYC

ID	Bioswale Area (sf)	Catchment Area (sf)	Bioswale Dimensions (ft)			Percentage of 1-Hour Storms Completely Retained					
			L	W	D	2 yr (1.35 in.)	5 yr (1.85 in.)	10 yr (2.35 in.)	25 yr (2.7 in.)	50 yr (3.1 in.)	100 yr (3.35 in.)
1	800	14091	100	8	2.0	100	73	58	50	44	41
2	600	5922	75	8	1.2	100	75	59	52	45	42
3	1200	6563	150	8	0.7	100	74	58	51	44	41
4	600	2422	75	8	0.5	93	68	53	46	40	37
5	800	5522	100	8	0.8	95	69	54	47	41	38
6	600	4253	75	8	0.9	100	76	60	52	45	42
7	400	1310	50	8	0.5	100	80	63	55	48	44

TABLE 4.4

1-Hour Storm Depths for Different Return Periods (Seattle, WA)

Percentage of 1-Hour Storms Completely Retained					
2 yr (0.416 in.)	5 yr (0.511 in.)	10 yr (0.582 in.)	25 yr (0.696 in.)	50 yr (0.788 in.)	100 yr (0.887 in.)

bioswales that were feasible, and the percentage of 1-hour precipitation events with between 2- and 100-year return periods that could theoretically be stored in each (i.e., we calculate what percentage of the volume of precipitation falling on the catchment area "fits" inside the bioswale volume). The depth of these precipitation events is also shown in the table. Although it is assumed that each bioswale is "empty" at the onset of rainfall, the estimate is conservative because the volume of water that can be stored in the soil pores, as well as the loss of stormwater to infiltration and evaporation during the period of wet weather are all assumed negligible in this calculation.

As shown in the tables, the performance of the seven bioswales in NYC varies from bioswale to bioswale, and with the return period of the precipitation event. Each of the bioswales captures 93–100% of the 2-year storm, but only 37–44% of the 100-year storm, with interim capture percentages for the other storms shown in the table. Due to differences in size and placement, there is a 5–16% difference in the volume of stormwater captures by different bioswales installed on the same street and subjected to the same precipitation event. Because of differences in precipitation depths, there can be 56–60% differences in performance of a single bioswale.

To illustrate how GI performance levels can vary from city to city, we computed the same performance statistic for identically sized bioswales subjected to Seattle, Washington, precipitation events of the same duration and return periods. Overall, although Seattle receives less annual precipitation than New York, precipitation events typically occur at lower intensities over longer periods. Table 4.4 lists the depth of precipitation associated with storms of identical return periods (2–100 years) to those shown for NYC in Table 4.3. It is noteworthy that, as sized, all seven bioswales would capture 100% of all the Seattle storm events.

This simple example illustrates the importance of contextualizing GI performance statistics with overall local precipitation conditions, event type, as well as site and design parameters. Because of the temporal variability in precipitation, event characteristics in any one place and the importance of interevent dry spell duration in determining system "recovery" (i.e., dryout) and available storage capacity, a more robust measure of performance would be based on continuous simulations considering stochastic precipitation ensembles generated from historical sets of data.

Human Factors

Even if specific GI practices can be designed to cost-effectively control precipitation quantities at the parcel scale, many of these systems would need to be implemented across a watershed to make GI effective overall. Because land tenure in urbanized watersheds is fragmented, with private property accounting for a significant percentage of the total area, wide-scale implementation of GI would be most effective if multiple property owners adopted GI over reasonable periods of time. Although the effect that GI technologies have on the urban hydrological cycle and ecosystem function is a climatologically driven, biogeophysical process, decisions related to where, when, and how these technologies are adopted are socially determined.

Little is known about the perceptions that urban residents have of GI (Hill, 2007), however, previous work by the authors in urban watersheds in East Coast cities indicates that opinions vary subtly from community to community. At the grassroots level, the adoption of GI technologies is often debated not from a stormwater management perspective, but rather in the context of larger, urban social and environmental justice and sustainability issues: Will GI lead to gentrification? Will it increase property values? Is it an appropriate use for vacant, blighted, or tax-delinquent urban land? Can the long-term maintenance of GI lead to creating green jobs? Will it improve air quality? Reduce heat islands? Lead to improved public health? Public opinion on such issues varies widely based on demographic, cultural, and other indices.

As an example, Figures 4.3 to 4.6 depicts the response of 300 owners of multifamily residential buildings in the Gowanus Canal (Brooklyn, New

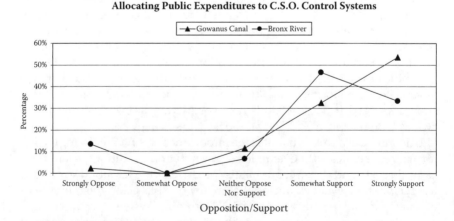

FIGURE 4.3
Allocating public expenditures to CSO control systems.

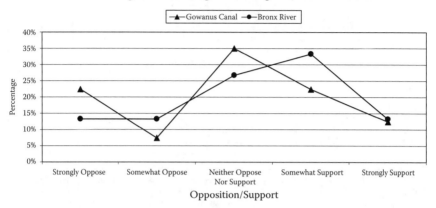

FIGURE 4.4
Installing a public underground storage tank to abate CSOs.

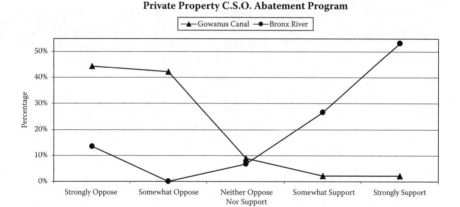

FIGURE 4.5
Private property CSO abatement program.

York) and Bronx River (Bronx, New York) watersheds who were mailed surveys that included questions about government programs to reduce combined sewer overflows (CSOs) using public (Figures 4.3 and 4.4) and private (Figures 4.5 and 4.6) property. Demographically, the Gowanus respondents fell within a higher economic bracket than their Bronx River counterparts; approximately 59% have an annual household income of over $95,000, whereas the Bronx proportion in the same bracket is only 15%. For racial differentiations, most Gowanus responders identified

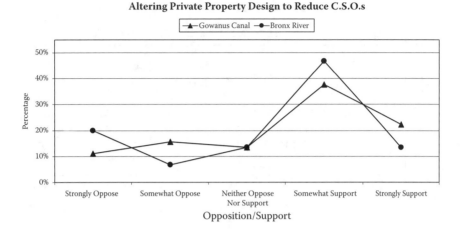

FIGURE 4.6
Altering private property design to reduce CSOs.

themselves as "White" (91.9%); whereas over half of Bronx responders (53.8%) identified themselves as "Hispanic," and 38.5% as "Black."

The response rates differed significantly between the two watersheds (17% in Gowanus versus 5% in the Bronx). Most respondents in both watersheds supported allocation of public expenditures to reduce CSOs, with about 10% more of the Bronx respondents supporting construction of a below-ground storage tank. The greatest difference in opinion was in the level of support for a program that uses private property to reduce CSOs, which achieved a significantly higher level of support among the Bronx River respondents. Interestingly, however, both samples assigned similar levels of support to alteration of the design of private property to reduce CSOs.

Accounting for stakeholder perspectives on GI adoption is an important step in evaluating the effectiveness of the GI approach to stormwater management. It requires that engineers and planners intimately engage with stakeholders in both the planning and implementation of GI (Novotny and Brown, 2007; Baker, 2009), and may involve incentive mechanisms such as those mentioned in Chapters 7 and 8 in this book.

Weaknesses in Conventional GI Program Analyses

Before most municipalities and water utilities commit to GI as an integral component of their stormwater management plans (simultaneously reducing their investment in other "gray" stormwater management [SWM] approaches), the various uncertainties described above should be explicitly considered. Decision makers can improve the design and reach of their programs if they better understand the likelihood that a public policy, incentive, or other mechanism promoting GI would yield a particular measurable outcome. As background to this overall approach, in this section we review traditional approaches to dealing with uncertainty in environmental decision making, and explain why they are not ideal for making decisions associated with GI.

Conventionally, infrastructure and other environmental decisions are made by teams of experts, with little input from the expected beneficiaries. Based on knowledge derived from similar projects, these experts recommend the "optimal" solution among several options that have been selected for consideration. Where the recommendations are based on measures of infrastructure performance, models and other forms of analysis are used to derive a single "expected outcome" among several alternative scenarios. Although this approach may provide the single best statistical estimate, it offers no information about the range of other possible outcomes and their associated probabilities. The problem becomes acute when uncertainty surrounding the forecast's underlying assumptions is significant, such as would be the case in a GI modeling problem.

In recognition of this problem, a common alternative is to create "high case" and "low case" scenarios to bracket the central estimate. This approach is problematic, however, because it gives no indication of the likelihood associated with the alternative outcomes. The "high case" may assume that most underlying assumptions deviate in one direction from their expected value (e.g., all GI installed will be low-cost, high-performance, and readily adopted), whereas the "low case" assumes those same assumptions simultaneously deviate in the opposite direction (e.g., high cost, low performance, not readily adopted). In reality, the likelihood that all underlying factors will shift in the same direction is just as remote as the case where they all assume their expected value.

Such analyses are often accompanied by a "sensitivity analysis" in which key forecast assumptions are varied one at a time to assess their relative impact on the expected outcome. As they are typically performed, sensitivity analyses do not address the problems outlined above inasmuch as (a) individual assumptions are often varied by arbitrary incremental amounts, and (b) in the real world, assumptions do not veer from outcomes one at a time. Rather, it is the total impact of simultaneous differences in the modeled assumptions that provide a practical perspective on the likelihood of a particular outcome.

All of these problems are further amplified when a particular decision involves assumptions about human behavior (e.g., likelihood that particular property owners will adopt GI). Although uncertainty in physical parameters can be characterized using various approaches (Singh, Jain, and Tyagi, 2007), human decisions are more difficult to fit into normative models. A wealth of empirical evidence (Kahneman, 2000) now challenges the notion that human responses (e.g., to a new GI policy or incentive) can be modeled as either uniform or rational processes. Much research (in particular, the so-called heuristics and biases literature) during the 1970s and 1980s demonstrated the divergence of descriptive accounts of human behavior from normative models. Indeed, many multipurpose water management plans developed "rationally" by "experts" in the United States during the early twentieth century are now considered failures because (a) the projects were eventually not needed, (b) the management plans included too many recommendations, and (c) the associated planning studies were expensive and time consuming (Hays, 1959 and White, 1969 as cited in Mukhatarov, 2008).

Beginning in the 1970s, pioneering efforts by the Army Corps of Engineers and others emphasized the benefits of engaging stakeholders, that is, those bearing the consequences of specific actions, in modeling and other decision-making processes. Today, public participation is now a required component of most large infrastructure projects (including those related to stormwater management). However, although it is agreed that planning efforts that focus solely on the "technical aspects of socio-technical systems" are inadequate (Ackoff, 1981 as cited in Elghali, et al. 2008), and that a participatory approach to planning yields projects that are eventually carried out with more success and less conflict than those born from top-down efforts, in practice public engagement is often nominal (Voinov and Bousquet, 2010). As it is typically approached, "public participation" is used as a means of obtaining buy-in to prescreened infrastructure scenarios of interest to a particular public agency. This kind of public "persuasion" is not likely to yield as significant levels of GI adoption as would be achieved when owners willingly and voluntarily choose to modify their property so that it captures more stormwater.

Proposed Methodology for Making GI Decisions

In this section, we propose a structured approach to making GI policy decisions. Specifically, we propose a five-step process for simulating the cost-effectiveness of community-supported GI scenarios. These five steps are:

1. Preplanning data gathering
2. Stormwater management goal setting
3. Stakeholder-facilitated GI scenario building
4. Cost-effectiveness simulations
5. Postmodeling policy development

Fundamental to the approach are probabilistic analysis techniques and a mediated-modeling process, both performed with the help of the low-impact development rapid assessment model (developed by a team including the authors). Version 2.0 of LIDRA is a web-based tool that can rapidly assess the cost-effectiveness of a wide-range of GI strategies and adoption rates to reduce annual runoff. In contrast to most expert-driven modeling processes, mediated modeling involves stakeholders as active participants in all stages of decision making, that is, from initial problem scoping to model development, implementation, and use (Costanza and Ruth, 1998; van den Belt, 2004). When attempting to simulate complex system processes, such as how human decisions might alter the flow of water through large urban watersheds, input from a broad range of stakeholders (residents, property owners, water utilities, local government, etc.) is required. The process of mediated modeling can help to build mutual understanding and even consensus among such stakeholders regarding, for example, the appropriateness of specific assumptions made during model construction and scenario analysis, and ultimately on a plan to move forward.

Overview of LIDRA 2.0

The low-impact development rapid assessment model was developed to enable GI cost-effectiveness calculations to be performed simply and rapidly through a larger mediated-modeling process. Simplicity, speed, and ease of use are all important characteristics of any tool used in a participatory process involving multiple stakeholders with different levels of technical expertise and understanding of the problem. Inserted into a larger GI decision-making process, LIDRA can be used to "test" community-derived GI scenarios from both a cost and performance perspective. LIDRA couples the expected life-cycle cost of a GI program with the effectiveness of specific GI technologies at reducing runoff, to arrive at the cost-effectiveness of GI as a runoff reduction strategy.

The first version of the model (v. 1.0) was a spreadsheet tool using a basic hydrologic simulation algorithm based on a typical rainfall year, and LCC computations made using discreet GI cost figures. Montalto et al. (2007) present that version of the model and demonstrate an application comparing the cost-effectiveness of GI and underground storage tanks as alternative means of controlling combined sewer overflows in New York City.

The current version of the model is a free web-based program with improvements that include: (1) a GIS interface allowing users to upload parcel and street data from small to large watersheds; (2) a means of specifying uncertainty in forecasts of GI adoption rates; (3) a more sophisticated (i.e., continuous) rainfall-runoff model driven by a stochastic precipitation generator operating at an hourly timestep; and (4) a means of specifying uncertainty in capital and annual costs using cost data from a national database. These enhancements enable the various forms of uncertainty to be quantified so that decisions can be based on the most realistic outlook on cost-effectiveness. The model is accessed using the LIDRA website (www.lidratool.org), which will be operational during the summer 2011.

A more thorough description of LIDRA 2.0 is provided elsewhere (Behr and Montalto, 2008; Yu et al., 2010; Aguayo et al., 2011; Yu et al., 2011). Below, we summarize several features of the program, before describing more thoroughly the entire five-step community-facilitated GI decision-making process.

Parcel and Street Data Upload

Users set up modeling "projects" composed of one or more "sheds." Each shed is composed of "parcels" and "streets." Parcels are broken up into subelements: roofs, yards, and driveways. Streets are broken up into subelements: driving lanes, parking lanes, sidewalks, curbside sidewalk regions, and intersections. Users can include as many different types of parcels and streets as desired. Data can be inserted manually (i.e., keyed in on the screen) or uploaded using a spreadsheet designed to interface with a typical GIS database. The latter would be used if the user had access to appropriate metadata, for example, as could be exported from a GIS-based tax parcel database.

A parcel or street "type" is defined uniquely by (a) the percentage of its area occupied by each of the different subelements, (b) the underlying soil type (sand, clay, or loam), (c) the specific GI strategies for which it is to be considered a candidate, and (d) the rate at which these GI strategies are expected to be implemented on the parcel or street over a 30-year time horizon.

Selection of GI Types, Locations, and Rates of Adoption

Users also specify GI scenarios that are relevant for the parcel and street types in the project. For parcels, users are requested to submit an "annual adoption rate" (e.g., what percentage of the total area of that type of parcel adopt GI measures each year). For street types, users specify a "repaving rate." The assumption is that the GI measures assigned to a particular street type will be incorporated only when the street is repaved. Ranges of values for these parameters can be derived from a facilitated charette process performed with local stakeholders, as described below.

Many different GI options can be assigned to each parcel and street type, therefore users are provided with a selection tree. Thirty different GI scenarios are provided for parcels. First, users must decide if the roof is

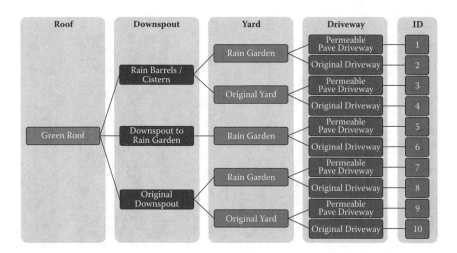

FIGURE 4.7
Parcel GI combinations assuming a green roof.

a candidate for a green roof, a blue roof (e.g., rooftop detention collar), or if no GI measures will be considered on the roof. After selecting the roof treatment, users make the other decisions shown in Figure 4.7, regarding treatments for the downspout, backyard, and driveway.

Similarly, users are presented with 16 different GI options for streets, depending on whether street trees are to be installed. Once users have made a decision about the use of street trees, they proceed by making the other decisions shown in Figure 4.8, for treatments to be considered for the sidewalk, curbside, and parking lane regions.

The GI scenario choices are stored in the relational database constructed using Microsoft SQL Server. The simulation spans a 30-year time span, and gradually phases in the GI measures according to the specified adoption/repaving rates. The water budgeting analysis (described next) computes annual runoff volumes from the model domain with and without the GI measures in place. A comparison of these values yields a time series representing the percentage of reduction in annual runoff (e.g., the effectiveness of the GI program) reported back to the user.

Stochastic Rainfall-Runoff Model

LIDRA does not simulate routing of stormwater through pipes. Rather it continuously tracks the cumulative volume of runoff generated from each parcel and street in the model domain, allowing users to compare the effects of various GI technologies implemented at different rates over 30 years (a time frame typical for infrastructure facility planning).

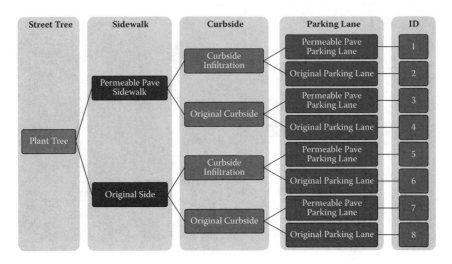

FIGURE 4.8
Street GI combinations considering addition of street trees.

The rainfall-runoff model is driven by stochastically generated precipitation ensembles that are derived from historical data provided by the user. Stochastic precipitation ensures that predicted levels of runoff reduction are realistic. Multiple precipitation realizations are generated using a nonparametric approach to retain the portability of the model to watersheds with different precipitation characteristics. Although details are beyond the scope of this text, in summary the synthetic precipitation time series are constructed by conditionally sampling historical values (a minimum of three years of hourly data is required) with replacement from a moving window, using a Markov chain process. The procedure, based on Lall et al. (1996), is used to generate 100 different realizations of local precipitation to generate annual runoff reduction possibilities.

A 30-year water balance is performed on each of the streets and parcels in the database with and without the GI measures in place. The water balance is performed by dividing each parcel type into three control volumes: the roof, the yard, and the driveway. The cumulative amount of runoff from the parcel over one timestep is computed in stages starting at the highest control volume (e.g., the roof). The total volume of runoff from the parcel is the sum of discharge to the drainage system from (a) the downspout, yard, and driveway, or (b) only the yard and driveway, as determined by the GI configuration specified by the user. A generalized control volume is shown below (Figure 4.9).

For subelements of parcels or streets comprising impervious surfaces, the water that is accumulated in surface "depressions" is tracked at hourly timesteps. Runoff is assumed to occur from the subelement only when

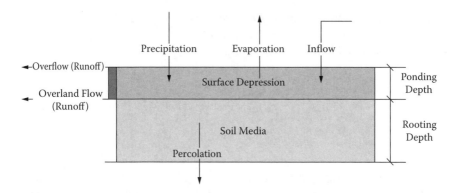

FIGURE 4.9
Generalized control volume used to compute runoff from each street and parcel surface in LIDRA.

the available storage in the surface depressions is exceeded. During dry spells, stored water is lost from the depressions due to regional evaporation rates, hardwired into the program.

For pervious subelements that consist of vegetation and soil, a modified version of the Thornthwaite Mather procedure is used to track the moisture content of the soil in the root zone at hourly timesteps. Runoff is assumed to occur from pervious surfaces only when the moisture content of the root zone exceeds saturation. When the moisture content exceeds field capacity, percolation to groundwater occurs. Between field capacity and the wilting point, the moisture content of the root zone is assumed to decrease due to evapotranspiration. The moisture content is not allowed to decrease below the wilting point. Direct precipitation and inflow from higher elevation subelements must exhaust the available pore capacity before runoff can occur.

GI Life-Cycle Costs, Program Costs, and Uncertainty

GI investments are assumed to be made over time across a watershed. Each GI system includes initial capital and long-term annual O&M costs. To compare GI investments and GI program scenarios, the GI LCC costs that occur in different periods are computed in present value (PV) terms using discounting. Present value costs reflect the total future costs of an investment or program if they were all accounted for in the present (and grew with the interest rate until they were expended). The PV of the program could also be annualized to represent the annual financial commitment of a program over time.

The equation below shows how the present value of a GI installation of type "i" in year 1 is computed using unit costs ($/Area) over a particular installation area. It is assumed that all capital costs are expended in the first year and average annual O&M costs are discounted using a real discount rate over a 30-year planning horizon.

$$PV^1_i = \left[IC_1 + \sum_{t=2}^{30} \left[AC_i \left(\frac{1}{(1+r)^{t-1}} \right) \right] \right] \times A_1$$

where:

PV_i^1 = present value of all GI measures of type "i" installed during year 1;

IC_1 = initial capital cost ($/area) to install all GI of type "i" during year 1;

AC_i = average annual O&M costs ($/area) to operate and maintain all GI of type "i" during each year during its useful life;

r = real discount rate (set at 5%);

A_1 = area of LID/GI of type "I" installed during year 1.

LIDRA accounts for uncertainty in cost and durability on capital costs, O&M costs, and life expectancy. LIDRA assumes the cost range is reflected by a probability distribution (e.g., triangular as a starting point) of the national-level cost survey data presented in Table 4.1. Annual real escalation of costs is included as a user-specified input. Future cost growth is assumed to be in real terms to be consistent with a real discount rate.

By quantifying ranges (i.e., probability distributions) for each cost parameter and performing Monte Carlo simulation, all cost inputs are permitted to vary simultaneously within their distributions, thus avoiding the problems inherent in conventional economic sensitivity or scenario analyses. The approach also recognizes interrelationships between variables and their associated probability distributions.

Five Steps of Implementation

In this section, we describe how LIDRA can be used in a larger mediated modeling process to help make GI program decisions. A municipality or water utility might begin such a process to avoid the potential pitfalls of expert-only planning processes. As laid out below, the process could be a collaborative stakeholder process between policy makers and community groups to identify a suitable GI program.

The process entails a series of structured steps involving community engagement with quantitative analysis. Fundamental to the process are the principles of transparency and inclusiveness. This process draws from the experience and practice of the authors and their affiliated firms and institutions that have extensive experience performing economic cost-benefit and life-cycle cost-effectiveness analyses, stakeholder-driven participatory planning efforts, and green infrastructure hydrological and hydraulic assessments. These processes engage participants to help identify and quantify sources of uncertainty.

In this exercise, community members, planners, and decision makers would all participate in the selection of GI scenarios to consider for different parcel and street types. Because LIDRA can be run rapidly and in real-time, many possible scenarios could be tested iteratively. In viewing and interpreting the multiple model outcomes, participants will gradually develop an understanding of the hydrological and cost dynamics underlying the cost-effectiveness calculations.

This communication helps to establish common understanding among the participants of what is and is not possible based on the best available information for community-specific applications. Community members would be able to provide insight directly into policy formation through their perspectives about which GI would be acceptable, how quickly it may be adopted, and whether the estimated costs warrant a dedicated program. Policy makers could use the dialogue and data to help assess if a community is suitable for a GI program and how the program should be structured to achieve intended results.

Step 1: Preplanning Data Gathering

The first step in the mediated modeling process would be initiated once GI has been identified as a potentially appropriate runoff reduction measure in a given area. A preplanning data gathering effort would:

- Select and geographically define the watershed of interest.
- Obtain relevant parcel, street, and soil data within the watershed.
- Obtain local hourly precipitation time series.
- Identify stakeholders and property owners with a potential interest in GI planning.

The project facilitators complete this step by creating a LIDRA project and uploading the relevant parcel, street, soil, and precipitation data.

Step 2: Stormwater Management Goal Setting

Next, a workshop is convened to discuss the feasibility of a GI program in the selected watershed. The workshop would entail invited presentations by community and government stakeholders on local perspectives and data on the impacts of runoff in the watershed. Local residents outside the selected watershed who have implemented GI could also be invited to present their experiences.

To stimulate discussion, the watershed scientists could perform hypothetical GI scenarios with LIDRA to illustrate results on a screen in real time. The sensitivity of results to adoption rates, costs, and performance could be presented to ensure that stakeholders are well informed about options and outcomes.

Step 3: Stakeholder-Facilitated GI Scenario Building

A second workshop meeting would be convened to discuss specifically which GI scenarios could most appropriately be considered on streets and parcels in the study area. The facilitators would prepare for this meeting by generating large-scale printouts of "prototype" parcel and street configurations. During the meeting, participants would be asked to match specific GI combinations with the prototypes in smaller breakout groups (Figure 4.10). These "charettes" can be accomplished with simple tracing paper laid out on top of poster-size printouts of the prototypes. Representatives from each of the breakout groups are then asked to present the leading GI configurations back to the general assembly (Figure 4.11). The moment of reporting is a potentially empowering one, during which individuals who do not traditionally have the opportunity to discuss local infrastructure and urban design issues, get the opportunity to address their peers in this capacity. These kinds of experiences contribute to developing respect and common understanding among different stakeholders, an invaluable accomplishment as the planning process moves forward. At the end of the session, participants would be asked to identify the overall most promising GI configurations and scenarios. These scenarios are used in Step 4.

FIGURE 4.10
A breakout "charette" in Philadelphia during which local stakeholders identify likely GI scenarios for prototype local parcel and street configurations.

FIGURE 4.11
The designs developed by the breakout groups are presented back to the plenary for a vote.

Step 4: Cost-Effectiveness Simulations

A more focused set of LIDRA simulations and analyses are performed on selected GI scenarios based on the workshop. A number of outputs are produced including cost-effectiveness contours and cost and performance forecasts. Each graphical presentation of data enables side-by-side comparisons of different GI scenarios. Graphical depictions of program performance (Figure 4.12) incorporate the uncertainty associated with precipitation and rate of adoption. Here the dotted lines represent the region of 90% confidence bracketing that uncertainty. The figures provide a perspective on the percentage of reduction in runoff that can be achieved over time. A wider band overall indicates larger uncertainty. Figure 4.13 is a sample output of the LCC curves accompanying a specific set of runoff reduction predictions, with the dotted lines representing the uncertainty associated with GI life-cycle costs and escalation factors.

Cost-effectiveness results such as those shown in Figure 4.14 indicate how to represent the uncertainty in program cost and effectiveness of a single GI system. This graphic is a joint distribution of total program cost (horizontal axis) and effectiveness (vertical axis) with the frequency (shown as percentiles) of any cost and effectiveness combination. This graphic provides a color-coded, two-dimensional perspective on the

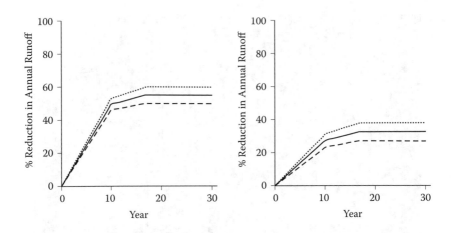

FIGURE 4.12
Side-by-side comparison of alternate GI scenarios emerging from the stakeholder-driven GI scenario-building process.

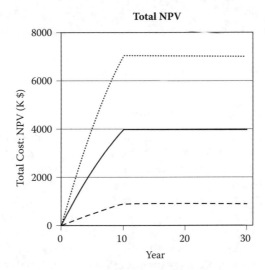

FIGURE 4.13
LIDRA cost curves presenting the NPV of a particular GI scenario.

vertical axis. Program cost represents the total present value costs of all individual GI systems implemented over time across the watershed. The effectiveness, measured in terms of total annual percent reduction in run-off is similarly evaluated for the entire GI build-out. Similar GI systems can be plotted in this manner to reflect their own variability factors.

Making decisions with these results is improved with an appreciation of the uncertainties. The most likely level of cost-effectiveness is a program

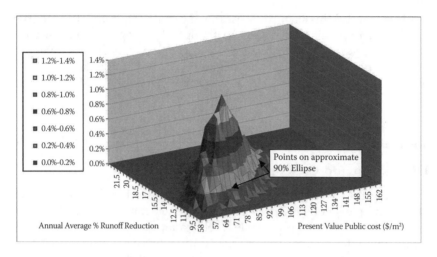

FIGURE 4.14
Example of cost-effectiveness computation with uncertainty.

that would cost around \$82/sq. meter and generate 12.5% reduction in annual runoff—shown at the peak of the joint distribution. But this is just one possible outcome. To interpret the fairly wide range of cost and effectiveness outcomes, bands are constructed around the joint distribution. An approximate 90% confidence interval ellipse includes all points within and above the light-colored band shown in Figure 4.14. This ellipse is a statistically computed region that includes all combinations of program cost and effectiveness, with a 90% level of confidence. For example, at a present value unit cost of \$85/sq. meter, there is a 90% level of confidence that the average annual reduction in runoff is between 10 and 17%.

A lower boundary of uncertainty tolerance can also be derived from this ellipse. This boundary is represented by a curve that extends from one end of the 90% ellipse, tracing the lowest percentages of reduction per unit cost on this ellipse, up to the other end of the ellipse across its long axis. All levels of percent reduction per unit cost above this curve would be achievable with a 95% level of confidence (noting that the lower ellipse drops off 5% on each side of the line). The curve trends upward because cost-effectiveness changes with increasing rates of adoption. The slope of this curve could be used to compare cost-effectiveness with other GI strategies and with conventional runoff reduction

Step 5: Postmodeling Policy Development

The final step is for the stakeholder group in partnership with local decision makers to interpret the modeling results and to agree on a GI policy.

It is important at this stage to assess the uncertainty in achieving program goals in a particular watershed. Although the program could achieve goals, the probability of achieving these goals may be low. Refinements could be performed to test the sensitivity of the results to changes in inputs (e.g., adoption rates or local cost adjustments).

Discussion and Conclusions

Projecting the extent to which GI can cost-effectively reduce urban runoff is a complex modeling problem involving extensive uncertainty in the cost, performance, and achievable level of adoption of GI. It is anticipated that as more GI systems are brought online, more extensive and useful datasets will become available. These include cost data, as well as adoption rates that have been achieved in different types of communities.

LIDRA is a simple approach to assessing the potential cost-effectiveness of GI as a means of reducing runoff. Implemented in the context of a mediated modeling effort, it can be used to integrate stakeholder perspectives into the decision-making processes aimed at devising a locally appropriate GI implementation program. Specifically, LIDRA allows the user to vary adoption rates and to compare different GI scenarios implemented on different parcel and street types, drawing GI costs from national datasets.

As these datasets become more robust, the uncertainty in LIDRA predictions will be reduced. A stochastic rainfall-generator incorporates uncertainty associated with local climatic conditions into the GI cost-effectiveness calculations. This is a significant improvement over event-driven hydrological simulations that do not take into consideration the variability in GI systems as it relates to storm characteristics and antecedent conditions. LIDRA is an improvement over alternative expert-driven modeling efforts based on best guesses of relevant physical, economic, and human behavioral characteristics.

As an SWM planning tool, LIDRA requires fewer input parameters than a more complicated distributed hydrological and hydraulic model of the catchment and sewer system such as the EPA Stormwater Management Model (SWMM), and can therefore be used to assess rapidly the potential cost-effectiveness of GI in a particular urban watershed. LIDRA can be used to evaluate quickly the potential for GI programs within and across watersheds. Within watersheds, this method could be used to quantify the GI implementation required to achieve significant decreases in runoff. Across watersheds, LIDRA can be used to prioritize municipal spending on SWM and compare the potential of individual GI systems to achieve water quality goals in different watersheds.

With the improvements incorporated into version 2, LIDRA takes advantage of modern analytical techniques to provide decision makers with the best available information on whether and where GI could be successful. Results include not only a most likely value of cost-effectiveness, but the probability that cost-effectiveness could be lower than expected. Use of the model is best undertaken in a transparent process in which experts and stakeholders provide input to the assessment of uncertainty. This process has been shown to be an effective way of communicating model assumptions, obtaining critical inputs, and fostering informed decision making.

References

Ackoff R.L. (1981). The art and science of mess management. *Interfaces*, 11:1, 20–26.

Aguayo, M., Z. Yu, M. Piasecki, and F.A. Montalto (2011). Development of a database and website for low impact development rapid assessment (LIDRA) tool version 2.0, J. Hydroinf. (in preparation).

Baker, L.A. (Ed.) (2009). *The Water Environment of Cities*. New York: Springer Science and Business Media.

Basinger, M., F.A. Montalto, and U. Lall. (2010). A rainwater harvesting system reliability model based on nonparametric stochastic rainfall generator. *J. Hydrol.*, 392:105–118.

Center for Neighborhood Technology (CNT) (2010). Accessed online at http://www.cnt.org/publications

Costanza, R., and M. Ruth. (1998). Using dynamic modeling to scope environmental problems and build consensus. *Environ. Manage.*, 22:183–195.

Elghali, L., R. Clift, K.G. Begg, and S. McLaren (2008). Decision support methodology for complex contexts. *Engrg. Sustain.*, 161:7–22.

Hill, K. (2007).Urban ecological design and urban ecology: An assessment of the state of current knowledge and a suggested research agenda. In V. Novoty and P. Brown (Eds.), *Cities of the Future: Towards Integrated Sustainable Water and Landscape Management*. Washington, DC: IWA.

International Stormwater BMP Database (2010). Accessed online at http://www.bmpdatabase.org/

Kahneman, D. (2000). A psychological point of view: Violations of rational rules as a diagnostic of mental processes. *Behav. Brain Sci.*, 23(5):681.

Lall, U., B. Rajagopalan, and D.G. Tarboton et al., (1996). A nonparametric wet/dry spell model for resampling daily precipitation. *Water Resources Res.* 32(9):2803–2823.

Montalto, F. A., C.T. Behr, K. Alfredo, M. Wolf, M. Arye, and M. Walsh. (2007) A rapid assessment of the cost effectiveness of low impact development for control of combined sewer overflows. *J. Landscape and Urban Planning*, 82:117-131.

Mukhatarov, F.G. (2008). Intellectual history and current status of integrated water resources management: A global perspective. In Pahl-Wostl et al. (Eds.) *Adaptive and Integrated Water Management: Coping with Complexity and Uncertainty.* New York: Springer.

Novotny V., and P. Brown. (2007). (Eds.), *Cities of the Future: Towards Integrated Sustainable Water and Landscape Management.* Washington, DC: IWA.

NYC. (2008). Sustainable Stormwater Management Plan. Accessed at http://www.nyc.gov/html/planyc2030/html/stormwater/stormwater.shtml

NYC. (2010). NYC Green Infrastructure Plan accessed online @ http://www.nyc.gov/html/dep/html/stormwater/nyc_green_infrastructure_plan.shtml

Philadelphia Water Department (2009). A triple bottom line assessment of traditional and green infrastructure options for controlling CSO events in Philadelphia's watersheds. Accessed online at : http://wiki.epa.gov/watershed2/index.php/A_Triple_Bottom_Line_Assessment_of   _Traditional_ and_Green_Infrastructure_Options_for_Controlling_CSO_Events_i n_Philadelphia's_Watersheds

Singh, V.O.P., S.K. Jain, and A. Tyagi. (2007). *Risk and Reliability Analysis: A Handbook for Civil and Environmental Engineers.* Reston VA: ASCE Press.

Smart Growth Network. (2010). Philadelphia outlines $1.6 billion green infrastructure plan for rainwater overflow" Accessed online @ http://www.smartgrowth.org/news/article.asp?art=7286&state=39

USEPA. (2007). Green infrastructure statement of intent. Downloaded July 2009 at http://cfpub.epa.gov/npdes/greeninfrastructure/information.cfm

USEPA. (2010). Green infrastructure main web page. Accessible online at http://cfpub.epa.gov/npdes/home.cfm?program_id=298

van den Belt, M. (2004). *Mediated Modeling: A System Dynamics Approach to Environmental Consensus Building,* Washington, DC: Island Press.

Voinov, A., and F. Bousquet. (2010). Modeling with stakeholders. *Environ. Model. Softw.,* 25:1268–1281.

Yu, Z., M. Aguayo, F.A. Montalto, and M. Piasecki. (2011). Precipitation synthesis and runoff reductions in LIDRA 2.0: Theoretical formulation and validation efforts. (in preparation).

Yu, Z., M. Aguayo, M. Piasecki, and F.A. Montalto. (2010). Developments in LIDRA 2.0: A planning level assessment of the cost-effectiveness of low impact development. *Proceedings of the ASCE Environment and Water Resources Institute Conference, Providence, RI, May 16–20, 2010.*

Zalewski, M. (2000). Ecohydrology—the scientific background to use ecosystem properties as management tools toward sustainability of water resources. *Ecol. Engrg.,* 16(1):1–8.

Zhou, J., F.A. Montalto, Z.K., Erdal, and P.T. Mcreanor. (2009). Sustainability. *Water Environ. Res. 2009 Lit. Rev.,* (39):1451–1489.

5

The Economics of Green Infrastructure and Low-Impact Development Practices

Ahjond S. Garmestani, Janet Clements, Joanna Pratt, and Lisa Hair

CONTENTS

Introduction

Managing stormwater runoff through traditional "gray" infrastructure systems (e.g., collecting large quantities of runoff for rapid disposal via pipes and storage tunnels) has resulted in a variety of challenges, including high costs associated with infrastructure construction, maintenance, and repair; very high costs for mitigating combined sewer overflows (CSOs); and continuously increasing pollution of surface waters across the country due to nonpoint source runoff from rural, suburban, and urban areas. These problems are exacerbated as population and development continue to increase and new challenges arise, such as increasing energy costs, expanding areas of environmental degradation due to urban sprawl, and aging water infrastructure. As these concerns converge, it is becoming clear to many cities, watershed management districts, and water utilities that a new approach to stormwater management, which focuses on sustainability and meets the needs of multiple stakeholders, will be required to help ensure that the nation can provide the quality and quantity of water demanded in the future.

Low-impact development (LID) is an approach to land development and redevelopment that incorporates sustainable stormwater management practices. LID activities prevent or reduce the impact of development on stormwater runoff and water quality through approaches such as infiltration, evapotranspiration, and rainwater use. Specific LID techniques include residential cluster development (instead of large lot subdivisions), open space preservation, reforestation, and "green infrastructure" (GI) options such as depressional infiltration areas or tree planting (termed "bioinfiltration"), permeable pavements, rain gardens, rain barrels and cisterns, green roofs, and vegetated swales. LID can result in a number of environmental, economic, and social benefits (i.e., the "triple bottom line," or TBL; Table 5.1). Communities throughout the United States are beginning to recognize these benefits, and have become increasingly interested in implementing LID-based approaches. However, because LID and GI have not yet been implemented on a broad-scale basis, a number of uncertainties surround the economics of these approaches compared to traditional gray infrastructure. For example:

TABLE 5.1

Examples of the Environmental, Economic, and Social Benefits of LID

- Environmental benefits
  - Improved water and air quality
  - Improved groundwater recharge
  - Improved dry-weather stream baseflow
  - Energy savings
  - Reduced greenhouse gas emissions
  - Reduced heat stress
  - Flood protection
  - Reduced sewer overflow
- Economic benefits
  - Reduced gray infrastructure construction and maintenance costs
  - Increased economic development
  - Green jobs
  - Increased land and property values
- Social benefits
  - Improved aesthetics
  - More urban greenways
  - Increased education of the public about their role in stormwater management

1. The capital and operations and maintenance (O&M) costs associated with LID- and GI-based technologies are very site-specific. Furthermore, technologies or approaches that work well in some communities may not be applicable in others. Thus, it is difficult to develop estimates or "rules of thumb" regarding costs.
2. Many entities have expressed concerns about the O&M costs associated with LID approaches, especially as compared to more traditional, gray infrastructure. There are limited data available on O&M costs for LID approaches.
3. There is uncertainty regarding the cost-effectiveness of LID approaches compared to traditional infrastructure.
4. The upfront capital costs associated with LID and GI are perceived to be more than those associated with traditional infrastructure.
5. It is difficult to monetize many of the nonmarket benefits associated with LID and GI (e.g., air quality, aesthetic value).

Given these uncertainties, it can be difficult for utilities and other implementing agencies (e.g., communities, municipalities, water management districts, utilities) to conduct a thorough economic analysis of LID-based options for stormwater management. However, as development pressures increase and aging infrastructure needs to be replaced or expanded, it is essential that agencies and municipalities are able to adequately compare and evaluate the economics of various alternatives (including both LID and more traditional, gray infrastructure approaches). To assist utilities and other agencies in conducting these types of analyses, the U.S. Environmental Protection Agency (EPA), Office of Watersheds, Oceans, and Wetlands, retained Stratus Consulting to develop a series of case studies of public entities in different parts of the United States that have conducted economic evaluations of their LID programs. As discussed in subsequent sections of this chapter, the case studies highlight various objectives and types of economic analyses, as well as a variety of LID-based programs. The following sections provide an overview of the approach used to select and develop the case studies, a description of the types of economic analysis conducted by the case study communities, an overview of the case studies, and a summary of key findings and lessons learned from the case studies.

Identification of Case Studies

As a first step to this research, Stratus Consulting (Stratus), under contract to the EPA, conducted a review of more than 45 communities with existing and planned LID- and GI-based programs. Communities included in this review were identified based on an extensive web search, information from recognized LID sources [e.g., the LID Center, the Water Environment Federation, and EPA GI (http://cfpub.epa.gov/npdes/home. cfm?program_id = 298) and LID (http://www.epa.gov/owow/nps/lid/ index.html) websites], interviews with EPA representatives from each of the 10 EPA regions, and personal knowledge and experience of the research team (including both EPA and Stratus team members). The project team developed a spreadsheet to track information collected for each community. To the extent possible, this included a description of the community's LID program and information on the economic analysis conducted (if any). Other relevant information, such as the size and geographic location of each entity, was also documented. Next, Stratus contacted the entities initially identified as potential case study candidates. Based on the information collected, the project team prioritized the list of potential case study entities for further review. After several rounds of additional information collection and discussions with the case study candidates, the project team identified nine case study communities for this study.

Although many of the candidates would have served as excellent case studies, the final nine communities were chosen to represent a variety of different types of economic analyses and LID programs, as well as broad geographic and demographic ranges. The final nine case study entities were:

1. City of Lenexa Public Works Department, Watershed Division, Lenexa, Kansas

2. Charlotte–Mecklenburg Storm Water Services, Mecklenburg County, North Carolina

3. Capital Region Watershed District, St. Paul, Minnesota

4. Mayor's Office of Long-Term Planning and Sustainability, New York City (NYC), New York

5. City of West Union, Iowa and the Iowa Department of Economic Development (IDED)

6. Alachua County Environmental Protection Department and Public Works Department, Alachua County, Florida

7. Bureau of Environmental Services (BES), Portland, Oregon

8. Sun Valley Watershed, Los Angeles County, California

9. Philadelphia Water Department (PWD), Philadelphia, Pennsylvania.

The project team conducted an in-depth interview with a representative from each of the case study communities. Based on these interviews, as well as reports and data provided by each entity, the team developed case studies highlighting each community's LID program; the role of the economic analysis conducted; the methods, results, and outcomes of the economic analysis; and key lessons learned in implementing the economic analysis or LID program. Much of the information in this study is taken directly from these interviews, rather than published sources, and information is provided here on the relevant project contacts.

Components and Types of Economic Analysis

A key finding of this research is that utilities and other implementing agencies have employed a number of different economic analysis techniques to evaluate alternatives for stormwater management. As evidenced in the case studies, economic analyses can range in complexity from a simple assessment of the capital costs of various alternatives to a comprehensive evaluation of the nonmarket benefits and costs of LID practices. The use of a specific technique or type of analysis depends on a variety of factors including the objective of the economic analysis (e.g., to compare costs of green and gray infrastructure solutions or communicate the value of GI benefits to the public and regulators), the budget available for conducting the analysis, and the type of LID program. The different types of economic analyses represented in the case studies include the following.

Capital Cost Assessment

An assessment of the capital costs of various alternatives can be used to compare the upfront costs associated with both LID and gray infrastructure alternatives. As detailed in the case study summarized below, the City of Lenexa, Kansas, used capital cost assessment to show savings associated with the implementation of LID- and GI-based best management practices (BMPs) compared to traditional development techniques.

Life-Cycle Assessment

Life-cycle costs are defined as the sum of the present value (PV) of investment costs, capital costs, installation costs, O&M costs, and replacement and disposal costs, over the lifetime of a project or program. Similarly, life-cycle benefits represent the PV of project benefits that accrue over its lifetime. Life-cycle net benefits reflect the PV benefits minus the PV costs of a project. Life-cycle net benefits are also known as the net present value (NPV) of the project. For a rural green street project, West Union calculated the life-cycle cost savings associated with the use of permeable pavement compared to traditional nonpermeable pavement. The analysis showed that although the permeable pavement would cost more up front, the city would realize annual savings beginning in year 15 of the project due to reduced O&M costs compared to traditional pavement.

Cost-Effectiveness Analysis

Communities can use cost-effectiveness analysis to more directly compare various LID and gray infrastructure solutions. Cost-effectiveness analysis is used to determine capital costs or life-cycle costs per unit of a specific measure. For example, Charlotte–Mecklenburg Storm Water Services evaluated the cost-effectiveness of LID solutions based on reduction in total suspended solids achieved per dollar spent. Severe stream erosion caused by traditional stormwater management in a developing watershed was the transportation of scoured sediment to the drinking water reservoir, an immediate problem that required prioritizing limited resources.

Benefit Valuation

Several entities are interested in quantifying the benefits provided by various alternatives or projects over time. For example, Alachua County, Florida quantified the increase in property values that occurred as a result of their LID open space preservation program, Alachua County Forever (ACF). This analysis was conducted in response to public concern over the property tax revenue loss associated with acquiring open space for preservation.

Benefit–Cost Analysis

Benefit–cost analysis is a common accounting framework used to evaluate the net effect of a proposed program or project (e.g., Do the benefits outweigh the costs? Who benefits? Who incurs the costs?). Los Angeles County, California (Sun Valley Watershed); Portland, Oregon; and Philadelphia, Pennsylvania conducted benefit–cost analyses of their LID-based programs. All three entities included triple bottom line (TBL) components in their analysis, as described below.

TBL Analysis

TBL analysis is best described as an expanded benefit–cost analysis that takes into account the financial, environmental, and social benefits and costs (i.e., the "triple bottom line") associated with a project or program. As seen in the Philadelphia case study, nonmarket benefits and costs are typically included in these types of analyses. TBL analysis provides a good communication tool for presenting benefits and costs across the three bottom lines.

Summary of Case Studies

This section provides a summary of the nine case studies, including a brief description of the various LID programs implemented by the case study entities, and an overview of the economic analyses conducted. The LID projects and programs implemented by the case study entities vary significantly. Factors influencing the implementation of specific LID or GI practices include the entity's needs and objectives (e.g., improve water quality, reduce stormwater runoff volume), geographic location (e.g., climate, dense urban areas versus rural or suburban areas), and budget, among other factors (Table 5.2). The following provides a brief overview of each case study entity's LID/GI program.

Public Works Department, City of Lenexa, Kansas: Demonstrating cost savings associated with new LID/GI development standards. The objectives of Lenexa's LID-based program, Rain to Recreation, include: (1) reduce flooding, (2) improve water quality and habitat, and (3) provide recreational opportunities. The program consists of both regulatory and nonregulatory approaches to stormwater management. Nonregulatory measures include major capital projects (e.g., lakes that serve as regional retention facilities), land acquisition, and stream restoration projects, as well as GI-based components such as green street improvements, rain gardens, bioretention areas, and wetlands. Specific regulatory measures include LID-oriented development standards for new development and an accompanying BMP manual, protection of priority natural resource areas, and a stream setback ordinance. Lenexa's economic analysis focuses on the impacts associated with adoption of the LID-based development standards.

Charlotte–Mecklenburg Storm Water Services, Mecklenburg County, North Carolina: Using cost-effectiveness to prioritize projects that reduce the impacts of rapid development that impair a drinking water reservoir. The primary objective of the Charlotte–Mecklenburg Storm Water Services LID-based stormwater program is to protect drinking water quality and water quality in general (e.g., recreation, habitat, endangered species). An increased volume of stormwater runoff (contributing pollutants) due to rapid development has been identified as the biggest threat to water quality in this region. The county's Capital Improvement Program has three primary LID-based focus areas: in-stream restoration, upland BMP retrofits (e.g., rain gardens, bioswales), and reforestation. In addition, the Town of Huntersville has implemented a postconstruction LID-based ordinance intended to mitigate water quality degradation from future development. Charlotte–Mecklenburg conducted a cost-effectiveness analysis that estimated the capital cost (in $) per pound of total suspended solids reduced.

TABLE 5.2

Summary of Case Study Entities

Entity	GI Program Description and Objectives	Role of Analysis	Type of Analysis	Key Metrics	Outcome of Analysis
Lenexa, KS	Adoption of LID standards/BMPs for new development and a systems development fee as part of *Rain to Recreation* Program. Program goals are: (1) reduce flooding, (2) improve water quality and habitat, and (3) increase recreation.	Evaluate impacts of development standards and fee prior to adoption Obtain stakeholder support	Capital cost assessment	Capital cost savings from implementation of BMPs compared to traditional development	Savings of tens to hundreds of thousands of dollars in site work and infrastructure costs associated with the application of LID/BMPs for different types of developments. In most cases, savings more than offset increases in costs due to the systems development charge. Analysis helped to gain developer support for adoption of the standards and fee.
Charlotte–Mecklenburg StormWater Services, NC	LID, stream restoration and other BMPs to reduce impacts of rapid development on drinking water quality.	Identify cost-effective means to reduce sediment loading to reservoir Prioritize projects	Cost effectiveness	Capital cost ($)/lb of sediment saved from entering stream	Watershed modeling analysis showed that LID was the only long-term approach to protect water quality. Stream restoration was found to be the most cost-effective means of immediately mitigating sediment loading to reservoir in this area. It is therefore the initial focus of the county's program. Prioritization allows county to implement need-based, rather than opportunity-based, projects.

Location	Description	Goals	Analysis	Metrics	Outcome
Capitol Region Watershed District, St. Paul, MN	Eighteen BMPs designed to reduce flooding and stormwater runoff, improve water quality, and enhance recreation in a local park	Identify low-cost stormwater solutions; Assess BMP cost-effectiveness; Guide future project development	Capital cost assessment; Cost effectiveness	Comparison of capital costs for GI vs. gray solution; Present value (PV) life-cycle costs for GI BMPs; PV cost/lb removal of pollutants; PV cost/cu ft stormwater reduction	Substantial cost savings and water quality benefits with GI compared to gray infrastructure. Analysis helped CRWD validate a watershed approach to water resource management and increased awareness/support for GI.
New York City, NY	Distributed GI controls to reduce stormwater runoff and CSOs, improve water quality, and increase public access to tributaries	Develop potential stormwater strategies; Prioritize pilot projects	Cost effectiveness	PV life-cycle costs; PV costs/gallon of runoff captured; Comparison of GI to gray infrastructure	Cost savings with GI compared to gray infrastructure. Analysis led the city to adopt 20 pilot projects, short-term strategies to supplement existing stormwater control efforts, medium-term strategies to develop cost-effective source controls, and long-term strategies to secure funding.
West Union, IA	Pilot community for Iowa Sustainable Green Streets Initiative due to need to replace aging infrastructure and reduce flooding in downtown	Gain support for project; Guide decision-making; Obtain funding	Life-cycle cost analysis; Benefit valuation (avoided costs)	PV life-cycle costs; Cumulative cost savings associated with LID compared to gray infrastructure	Lower maintenance and repair costs for porous pavements result in cumulative savings of $2.5 million. Analysis helped West Union secure funding. Without it, gray infrastructure approach would have been implemented

continued

TABLE 5.2 (Continued)

Summary of Case Study Entities

Entity	GI Program Description and Objectives	Role of Analysis	Type of Analysis	Key Metrics	Outcome of Analysis
Alachua County, FL	County acquires and preserves open space lands through Alachua County Forever Program (ACF) to reduce development impacts and improve water quality	Demonstrate benefits of ACF program to alleviate public concerns that ACF reduces property tax revenue	Benefit valuation	Increase in property values	Proximity to open space adds $8,000 to $10,000 to parcel value on average (and up to $25,000 per parcel), for a total impact of $150 million. This results in additional property tax revenues of $3.5 million per year.
Portland, OR	Ecoroof Program includes incentives for green roofs on privately owned buildings and green roof requirements for new, city-owned buildings	Gain program support Increase implementation of ecoroofs in the city	Benefit-cost analysis	PV life-cycle costs Avoided costs Environmental and social benefits Net benefits	Determined ecoroofs generate significant public and environmental benefits as well as benefits to developers and building owners (due to extended life of ecoroofs compared to traditional). Documenting benefits has shown the value of providing incentives and encouraged development of ecoroofs.

Sun Valley / LA County, CA	Goal of stormwater program is to alleviate flooding while providing multiple benefits. Program includes 15 projects with GI/LID components.	Demonstrate that although GI had higher costs than traditional infrastructure in this setting, the benefits of GI were significantly higher	Benefit-cost analysis	PV costs for capital, land, and O&M Environmental and social benefits B-C ratio	Demonstrated potential for multi-objective stormwater strategies to provide greater community value than a single objective flood control strategy. By quantifying benefits, LACDPW has engaged a wide range of agencies and stakeholders who might not otherwise have participated and/or provided funding for program.
Philadelphia, PA	Green City Clean Waters Program aims to reduce CSOs and improve water quality in part through distributed GI controls and comprehensive stream restoration program.	Demonstrate full range of societal benefits of GI to regulators and public	Triple bottom line analysis	Net PV life-cycle costs and benefits of GI and gray approaches Social, environmental, and financial benefits	LID-based approaches provide important environmental and social benefits, and these benefits are not generally provided by gray infrastructure. Analysis helped PWD determine that a GI-based approach, coupled with targeted gray infrastructure, is the best approach for city to follow.

Capitol Region Watershed District, St. Paul, Minnesota: Realizing cost savings and environmental benefits by using green stormwater infrastructure retrofits. The highly developed nature of the Capitol Region Watershed District (CRWD) leaves little flexibility for stormwater management. When CRWD evaluates BMPs, it is therefore primarily concerned with identifying areas to retrofit. Through its Arlington Pascal Stormwater Improvement Project (APSIP), CRWD has worked to reduce stormwater runoff volume and to improve water quality by reducing the amount of phosphorus, bacteria, mercury, nutrients, polychlorinated biphenyls, and turbidity discharging into Como Lake and the Mississippi River. APSIP consists of 18 stormwater BMPs, including eight rain gardens to address volume control and water quality; eight underground (under street) infiltration trenches to address stormwater rate and volume control; a large underground infiltration/storage facility to reduce flooding and improve water quality; and a regional stormwater pond, located on Como Park Golf Course. CRWD's economic analysis focused on comparing the capital costs of GI and gray options, and conducting cost-effectiveness analyses based on reductions in total phosphorus and suspended solids in local runoff. CRWD also evaluated the cost of various options per cubic foot of runoff reduced.

Mayor's Office of Long-Term Planning and Sustainability, New York City, New York: Bringing together agency stakeholders to assess the cost-effectiveness and feasibility of sustainable stormwater management in CSO areas. New York City (NYC) has developed a Sustainable Stormwater Management Plan as part of the city's broader sustainability initiative, PlaNYC. The overall water quality goal of PlaNYC is to improve public access to (and recreational use of) the city's tributaries from 48% today to 90% by 2030. Toward this end, the Sustainable Stormwater Management Plan evaluates the feasibility of various policies that, when fully implemented, will create a network of decentralized source controls to detain or capture over one billion additional gallons of stormwater annually. Essentially, the plan creates a strategy for increasing the use of GI throughout the city for stormwater management. The plan includes a variety of technological (i.e., structural) and nontechnological (i.e., nonstructural) source control measures related to four specific program areas: the public right of way, city-owned property, open space, and private development. Technological source control measures include green roofs, blue roofs, rainwater harvesting, vegetated controls, tree planting, permeable pavements, and engineered wetlands. Non-technological measures include design guidelines, performance measures, zoning requirements, and economic incentives. The city conducted cost-effectiveness analysis to compare the cost per gallon of runoff captured for GI and gray infrastructure alternatives.

West Union, Iowa: Long-term cost savings plus environmental and social benefits envisioned in rural Green Street Pilot Project. In partnership with the Iowa Department of Economic Development (IDED), West Union has developed an integrated approach to community sustainability and livability through the Iowa Green Streets Pilot Project, which includes the complete renovation of six downtown blocks. Primary objectives of the project include citizen safety, replacing aging infrastructure, improving water quality and habitat in a nearby trout stream, and reducing flooding in the downtown area. The project involves a number of LID/GI techniques including a permeable paver system (for the roadway and sidewalks), rain gardens, and biofiltration areas. West Union's economic analysis focused on a life-cycle cost analysis and estimation of cumulative cost savings associated with its Green Streets project.

Alachua County Environmental Protection Department and Public Works Department, Alachua County, Florida: Preserving suburban lands to improve water quality provides a good return on investment for the community. Alachua County developed its LID/GI-based program to help mitigate the impacts of past land development, and to plan for expected growth and associated impacts. The county's program includes development standards that require and provide incentives for the use of GI on public and private lands. Types of LID/GI approaches encouraged under the development standards include enhanced stormwater pond designs, permeable pavement, vegetated swales and rain gardens, cisterns and rain barrels, underground tanks, depressional parking islands and road medians, and permeable parking areas. The second component of the county's GI program is the Alachua County Forever (ACF) program. Through ACF, the county acquires, protects, and manages environmentally significant lands in order to protect water resources, wildlife habitat, and natural areas suitable for resource-based recreation. In addition, Alachua County conducted an analysis of the benefits (in terms of increased property values) of the ACF program.

Bureau of Environmental Services, Portland, Oregon: A cost-benefit analysis provides a basis for incentivizing ecoroof construction. The Portland Bureau of Environmental Services (BES) has developed a stormwater management program that recognizes the need for the promotion of sustainable stormwater management systems throughout the entire city. This stormwater initiative was driven by a need to address CSO compliance. The LID-based program helps the city meet its Municipal Separate Storm Sewer System (MS4) Discharge Permit, address CSO events, maintain water quality, and control flooding. BES conducted a cost-benefit evaluation of ecoroofs, which are one of the many citywide solutions initiated by BES. Additional LID program components include green streets, sustainable stormwater BMP monitoring, school BMPs, and a financing program.

Sun Valley Watershed, Los Angeles County Department of Public Works, Los Angeles County, California: Evaluating the benefits of using green infrastructure to reduce flooding. As part of the Sun Valley Watershed Management Plan, the Los Angeles County Department of Public Works (LACDPW) has developed a comprehensive LID-based program that offers a multipurpose approach to stormwater management. The program was developed to respond to the need to integrate flood control, stormwater pollution reduction, and water conservation (e.g., through infiltration and stormwater recycling) efforts. Program components also address additional community issues, such as the lack of recreational resources, wildlife habitat, and aesthetic amenities in the watershed. Sample projects include infiltration basins (e.g., Sun Valley Park Project), constructed wetlands, tree planting, development of parks and open space, and storm drains designed to convey stormwater to the project areas. LACDPW conducted a benefit–cost analysis to demonstrate the multiple benefits of its multi-objective stormwater strategies, and also included an analysis of TBL components.

Philadelphia Water Department, Philadelphia, Pennsylvania: Applying a TBL analysis to compare combined sewer overflow control options. The Philadelphia Water Department (PWD) is committed to development of a balanced "Land–Water–Infrastructure" approach to achieve its watershed management and CSO control goals. This method includes traditional infrastructure-based approaches, as well as a range of land-based stormwater management techniques and physical reconstruction of aquatic habitats. Land-based approaches include disconnection of impervious cover, bioretention, subsurface storage and infiltration, green roofs, swales, green streets (including permeable pavement), and tree planting. Water-based approaches include bed and bank stabilization and reconstruction, aquatic habitat creation, plunge pool removal, improvement of fish passage, and floodplain reconnection. Philadelphia's TBL analysis provided a comprehensive assessment of the financial, environmental, and social costs and benefits of its LID and gray infrastructure alternatives.

Summary of Key Findings and Lessons Learned

A variety of economic analysis techniques can be used to effectively evaluate LID-based approaches for stormwater management. The economic analyses employed by each case study entity are based on different needs and objectives, and pertain to very different LID programs. The key findings associated with the economic analyses conducted by the nine case study communities are summarized below.

Alleviate public concerns: In response to public concern regarding the loss of property tax revenues associated with preservation of open space lands, Alachua County, Florida, evaluated the increase in property values resulting from additional green space. The county completed a regression analysis on real estate sales that demonstrated the increase in land values for properties adjacent to open space more than offsets the revenue loss associated with acquiring open space for preservation. To alleviate potential concerns from the development community, Lenexa, Kansas conducted an economic analysis to evaluate cost savings associated with the use of LID-based techniques compared to traditional development practices.

Identify economically feasible stormwater management alternatives and guide future development of these approaches: CRWD and Charlotte–Mecklenburg Storm Water Services both conducted a cost-effectiveness analysis to identify the most economically feasible alternatives for stormwater management within their service area. Even without quantifying the multiple benefits associated with LID approaches, both entities found LID to be cost-effective when compared to more traditional solutions, and decided to implement LID approaches in future development. The objective of LACDPW's analysis was to evaluate the costs and benefits, including nonmarket benefits (e.g., air quality, water quality), of four stormwater management alternatives. This analysis allowed LACDPW and project partners to compare multi-objective solutions to the implementation of a more traditional, gray infrastructure approach to stormwater management.

Prioritize LID/GI activities to maximize benefit for minimum cost: NYC completed a first-tier cost-effectiveness analysis of feasible alternatives to prioritize specific LID/GI practices for pilot projects and further study. Philadelphia evaluated the net TBL benefits associated with various CSO control alternatives. The TBL analysis was initiated to demonstrate to regulators the magnitude of benefits associated with LID-based approaches.

Analyses have demonstrated the feasibility and cost-effectiveness of LID/GI-based approaches. Many entities have conducted economic analyses to evaluate the cost-effectiveness and feasibility of implementing various LID-based approaches. The results of these analyses have encouraged utilities to prioritize and implement LID approaches. For example, in evaluating cost-effectiveness, Charlotte–Mecklenburg Storm Water Services initiated LID approaches because model results indicated LID was the only approach that would achieve sufficient pollutant removal and prevent further degradation of the county's waterways. The county's cost-effectiveness analysis showed that stream restoration is the most cost-effective means of immediately mitigating sediment loading from scoured stream beds in this area. NYC's Sustainable Stormwater Management Plan is a comprehensive analysis of the feasibility and cost-effectiveness of stormwater management alternatives. NYC found that its proposed LID strategies—including sidewalk standards, road reconstruction standards, green roadway infrastructure, and requirements and incentives for low- and medium-density residences and other existing buildings—present significant opportunities for cost-effectively controlling stormwater and reducing CSOs. Based on the analysis conducted as part of its Sustainable Stormwater Management Plan, the city has developed and prioritized a series of promising stormwater strategies and pilot projects. The pilot projects provide a framework for further testing, assessing, and implementing decentralized source controls.

Other analyses demonstrate significant cost savings or higher net benefits compared to gray infrastructure approaches. Many of the case studies directly compared the costs of LID-based approaches to those associated with more traditional, gray infrastructure technologies. For example, The City of Lenexa found substantial cost savings associated with implementation of LID/GI-oriented BMPs for multifamily, commercial, and warehouse developments as compared to traditional approaches. West Union compared the life-cycle costs (including capital and O&M costs) associated with the use of a permeable paver system in the downtown area versus using traditional bituminous or Portland cement concrete pavement. Results of the analysis showed that although the use of porous pavement will initially be more expensive, the lower maintenance and repair costs will result in cost savings in the long run. The city will begin to realize these cost savings by year 15 of the project. Estimated cumulative savings (over a 57-year analysis period) are expected to amount to about $2.5 million. CRWD found significant capital cost savings compared to gray infrastructure. A new storm sewer for conveying untreated frequent floodwaters to Lake Como was estimated to cost $2.5 million compared to $2.0 million for implementing GI BMPs. In the Sun Valley watershed, LACDPW found that although the LID/GI-oriented retrofits selected for implementation in this highly urban area cost more than the traditional stormwater approach, the LID/GI solutions yield a much higher benefit to cost ratio. In Philadelphia, PWD performed a full comparison of green versus gray infrastructure to evaluate the best approach for investing the city's funds to solve the CSO problem in a dense urban environment. PWD's TBL analysis demonstrated that, in Philadelphia, for equal investment amounts and similar overflow volume reductions, supplementing gray infrastructure with GI provides 20 times the benefits (e.g., economic value associated with recreational opportunities, air quality, ecosystem enhancement, and other areas) compared to a single-purpose investment in traditional stormwater infrastructure. In Portland, the BES calculated the NPV of its ecoroof program to the public (i.e., the public stormwater system and the environment) and to private property owners (e.g., developers and building owners). Based on this analysis, BES concluded that construction of ecoroofs provides both an immediate and a long-term benefit to the public; the net present benefit is $101,660 at year 5, and $191,421 at year 40. For building owners, the benefits of ecoroofs do not exceed the costs until year 20, when conventional roofs require replacement. This finding showed that incentives for green roof implementation would provide more benefits to the public in the long term. Over the 40-year life of an ecoroof, the net present benefit of ecoroofs to private stakeholders is more than $400,000.

Although difficult to quantify, in cases where benefits have been evaluated, the multiple benefits of LID are much larger than those associated with gray infrastructure. All of the case study entities recognize the importance of the multiple benefits associated with LID and GI. In Portland, Philadelphia, and the Sun Valley Watershed, the case study entities were able to quantify and monetize benefits based on nonmarket or avoided cost economic valuation techniques. Some entities identified the key benefits of LID in their analysis but did not assign a value to these benefits. Several case study entities recognized the importance of the multiple benefits of LID and indicated that they hope to be able to quantify these benefits in the future.

For example, LACDPW monetized the multiple benefits associated with their proposed LID-based alternatives, including benefits associated with water conservation, recreational opportunities, improved community aesthetics, increased wildlife habitat, and reduced stormwater pollution. LACDPW found that by looking at the benefits per unit cost, rather than just lowest capital cost, it was able to provide a solution with more long-term value to the community, and by leading as an example, to the greater region at large. The Portland BES identified or quantified and monetized a wide range of benefits from ecoroof construction. Key benefits monetized in the city's cost–benefit analysis included (1) public benefits of reduced stormwater system management costs, habitat creation, improved air quality, and reduced carbon emissions; and (2) private benefits to developers and building owners of stormwater volume reduction, reduced energy demand for heating and cooling, avoided stormwater facility costs, increased roof longevity, and reduced cost due to heating, ventilating, and air-conditioning equipment sizing. PWD's TBL analysis of the financial, social, and environmental benefits associated with LID approaches in Philadelphia provided insight into the wide array of benefits of GI for urban residents and helped guide the city to obtain the best value for its residents. For each CSO control option, PWD quantified and monetized benefits associated with increased recreational opportunities, air quality improvements, water quality and ecosystem enhancement, creation of LID-based jobs, increased property values, and reduced urban heat stress. CRWD and West Union did not quantify the nonmarket benefits associated with their LID approaches, but did recognize their important role. CRWD recognized that the only benefit of the $2.5 million gray infrastructure alternative for stormwater management would have been to reduce localized flooding in Como Park. The lower-cost GI/LID option not only reduced flooding but also (1) reduced the volume of stormwater runoff, enhancing groundwater supplies, (2) improved water quality in an impaired recreational lake, and (3) enhanced the recreational amenities in Como Park. In its life-cycle cost analysis of permeable versus traditional pavement systems, West Union recognized, but did not quantify, the benefits of porous pavement, such as improved water quality, increased stream health and appearance, reduced storm sewer infrastructure and maintenance, improved pavement surface temperatures, and improved street appearance.

Economic analyses can help gain support for LID-based programs and projects. Many of the analyses described above have been used by the case study entities to gain stakeholder, regulatory, and/or public support for their LID programs. For example, prior to adoption of LID-oriented development standards and a systems development fee for new development, the City of Lenexa, Kansas analyzed potential impacts for different types of developments (e.g., residential, multifamily, commercial). As noted above, the analysis showed substantial cost savings associated with implementation of LID/GI-oriented BMPs compared to traditional development approaches. As a result of the analysis, the Lenexa City Council adopted the development standards and an accompanying BMP manual. In addition, the city gained developer support for the adoption of the ordinance and the systems development fee, which was also adopted. CRWD brought together three cities and a county to deal cooperatively with issues associated with excess runoff from impervious surfaces coming from their jurisdictions. Highlighting the results of the economic analysis for the planned BMPs was key to achieving awareness and support for these alternative approaches to stormwater management from the public and other stakeholders, and to developing the successful partnership. West Union reports that the results of its economic analysis played an important role in gaining public and city council support for the Green Street Pilot Project. The analysis also helped the city obtain financial support for the project, as granting agencies were able to evaluate the positive economic aspects of the program as part of the grant application and review process. By quantifying and monetizing the benefits associated with the proposed LID-based program for the Sun Valley Watershed, LACDPW has been able to engage support, including financial assistance, from a wide range of agencies and stakeholders who might not otherwise have been interested in participating in the project or providing funding. The Portland City Council has adopted a Green Building Policy that requires construction of an ecoroof for all new city-owned facilities and roof replacement projects (when technically feasible) as well as an incentive that offers developers floor area bonuses for buildings constructed with ecoroofs. The main purpose of the Portland BES cost-benefit analysis of ecoroofs was to provide further support for these programs and to further encourage the construction of ecoroofs in the city.

As LID implementation continues to increase, the economics of LID will be easier to evaluate and quantify. While conducting this research, the project team found that although many entities have begun to implement LID, very few have conducted economic analyses of their existing or proposed programs. This is largely due to uncertainties surrounding costs, operating and maintenance requirements, and difficulties associated with quantifying many of the benefits provided by LID. The case study economic analyses, as well as future analyses conducted as LID begins to be implemented on a wider scale, will help to serve as a model or guide for public agencies as they evaluate stormwater management investment approaches.

Lessons Learned as Reported by Each Community

As noted throughout the case studies, the use of economic analysis provided the case study entities with valuable insights, and also resulted in a number of "lessons learned." For example, although each entity conducted a thorough economic analysis to meet its specific goals and objectives, many entities indicated they would like to improve or expand these analyses. The biggest challenge mentioned by the case study entities involved developing estimates of the costs and benefits of their LID and GI actions. The case study entities indicated that estimating life-cycle costs, O&M costs, and monetizing the benefits of LID practices were particularly challenging. They also mentioned the need for "best practices" examples of how other agencies have successfully conducted economic analyses of their stormwater management programs.

Another key lesson learned by the case study entities is that it is important to be open and transparent in terms of how their economic analyses are conducted and the underlying assumptions, biases, and uncertainties related to calculating benefits and costs. This transparency, along with publicizing the estimated benefits of LID/GI activities, can help in gaining stakeholder support for alternative stormwater management programs. The remainder of this section highlights some of the challenges, lessons learned, and key insights related to each of the case study entity's economic analysis and how they used this analysis.

Lenexa, Kansas

Lenexa's *Rain to Recreation* program is highly supported by the city's residents and serves as a model for cities throughout the country. This is largely because the city has been open and transparent in developing its program and has included stakeholders at various levels of decision making. As Lenexa's program has progressed, monies and policies put in place have been related to compliance. The city council is interested in the life-cycle costs associated with these efforts and would like more information on these costs in the future.

Charlotte–Mecklenburg Storm Water Services, North Carolina

Charlotte–Mecklenburg Storm Water Services indicated that it would like to obtain better estimates of the O&M costs associated with different types of projects. This is particularly important because in some jurisdictions, the county is beginning to take over the maintenance of projects that were constructed by others. This often occurs, for example, in situations where a homeowner association might not be able to fund the necessary maintenance activities or inspection of LID projects within its development. With a better idea of the costs associated with maintaining different projects, the county can modify design standards to minimize these costs. A key outcome of Charlotte–Mecklenburg Storm Water Services' analysis is that the county has begun to shift from implementing opportunity-based projects to implementing need-based projects. This includes looking at a number of drivers for implementation and selecting projects that will provide the largest benefits.

Capitol Region Watershed District, Minnesota

CRWD developed the approach for their cost-effectiveness analysis without the benefit of having many examples to follow. They decided to prioritize their projects based on a few key questions that would help move their program forward (e.g., What are the costs? Are these techniques effective? What is the balance between construction and O&M costs?). One difficulty they encountered was being able to account for irregular maintenance costs (e.g., Como Park Pond will need to be dredged and infiltration trenches will need to be cleaned out, but it is difficult to determine how frequently these tasks will need to occur). To account for these costs, the district amortized (annualized) O&M costs over the project evaluation period (30 years). In most years, CRWD will spend a lot less than the estimated annual maintenance costs. CRWD plans to continue to conduct similar types of analyses on a somewhat regular basis (e.g., every two to three years). In subsequent studies, CRWD would like to quantify some of the secondary benefits of some of the BMPs (e.g., carbon footprint reduction, reduction in urban heat island effect).

New York City

NYC is pleased with the outcome of its Sustainable Stormwater Management Plan in terms of the economic analysis. Overall, although the plan indicates that there is tremendous potential for GI in NYC, further testing is needed. In the future, the city would like to look into the costs and benefits of GI at the watershed level rather than from a citywide perspective. A key challenge for NYC's program will be to coordinate successfully the various agencies and stakeholders that are involved in implementing the plan. Many of these agencies are supportive of LID/ GI, but are concerned about the O&M costs associated with this type of infrastructure.

West Union, Iowa

For West Union, a key challenge in putting together the economic analysis was being able to compare upfront and long-term costs for LID-based practices. Urban-based analyses do not transfer very well to smaller, rural community situations. For example, West Union compared the capital and O&M costs of the permeable paver system with those associated with traditional types of pavement. However, West Union does not have a regular maintenance plan for its infrastructure. Thus, the city will not realize cost savings if they are not maintaining the roads in the first place. West Union worked with the best information available, based on local experience of the contracted engineering firm and the city's public works staff. Nevertheless, they found that there was a lack of existing data for evaluating the costs of newer innovative practices, and that it was essential to make a number of assumptions.

Alachua County, Florida

Alachua County indicated that in hindsight, it would have liked to have focused its study on the property value impacts of the open space lands acquired by the county under ACF, rather than all open space properties in the county. The county has initiated a follow-up study to evaluate this impact. The county understands the importance of getting the message to its stakeholders that GI has a good return on investment (i.e., larger net benefits) as compared to traditional infrastructure, and plans to continue its efforts to educate the community about the effectiveness of GI implementation. The county would like to improve its education program by being able to put a value on its GI services (e.g., in terms of carbon sequestration, quality of life, improved health, and water resources value) and providing more examples of how the LID component of its development standards reduces the impacts associated with stormwater volume.

Portland, Oregon's Ecoroof Program

Portland BES has been able to quantify many of the cobenefits of ecoroofs, and found that publicizing these benefits (e.g., demonstrating the benefits that would accrue to the public by incentivizing the construction of privately owned ecoroofs) presents a more convincing argument for the program than simply describing the importance of stormwater management. Constraints faced by the BES include difficulties in extrapolating findings from the literature to the City of Portland and monetizing benefits. The BES hopes to monetize additional benefits (e.g., from carbon sequestration, reduced heat island effects, and increased habitats) in future studies.

Sun Valley Watershed, California

LACDPW plans to continue to conduct benefit–cost analyses on a project-by-project basis. The analysis presented in the Sun Valley Watershed Management Plan was completed in 2004 and costs have increased since then. More detailed analyses will be conducted for each planned project in order to evaluate and justify implementation. In addition, LACDPW reports that some projects will cost much more than originally anticipated, primarily due to unforeseen costs associated with private land acquisition. Given the county's current financial situation, cost increases for some projects may affect the ability to implement others. Future projects may be able to be designed to capture more stormwater in order to reduce requirements for other locations. Keeping up with maintenance requirements has been difficult for LACDPW, including maintenance associated with water quality monitoring of the extensive high-tech treatment units and groundwater monitoring systems, and cleaning of underground treatment systems. LACDPW may try to change the design of future projects to reduce maintenance needs. The plan does not take into account most of the maintenance costs associated with current projects, and there is currently no plan to evaluate these costs. Finally, LACDPW understands the importance of involving the community and other stakeholders at all levels of project design and implementation, including the economic analysis. It has worked with an environmental nonprofit organization, TreePeople, to educate stakeholders at community events, and has developed a project website that is actively used and frequently updated.

Philadelphia, Pennsylvania

PWD recognized that analyses of social and environmental benefits invariably require the use of assumptions and approaches that interject uncertainty about the accuracy or comprehensiveness of the empirical results. However, the methods and principles employed in PWD's TBL analysis are well-established approaches to quantifying benefits in the field of environmental and natural resource economics. Throughout its analysis, PWD attempted to be explicit and reasonable about the assumptions and approaches adopted. For transparency, the research team identified key omissions, biases, and uncertainties embedded in the analysis, and described how the results of the analysis would likely have been affected (e.g., whether benefits would have increased, decreased, or changed in an uncertain direction) if the omission or data limitation had been avoidable. In conjunction with these issues, a series of sensitivity analyses was conducted to explore how changing some of the key assumptions would affect findings.

Conclusion

This study of the economics of green infrastructure and low-impact development practices has revealed a number of conclusions about the feasibility of conducting economic analyses of LID and GI programs and the variety of ways that local entities have used these analyses. These conclusions, which have been illustrated in the preceding sections, are as follows.

The nine case study entities have conducted a variety of economic analysis methods to assess the feasibility and economic viability of LID and GI options as compared to traditional gray infrastructure. These analyses ranged from simple capital cost assessments to complex benefit–cost and TBL analyses. The case study utilities selected the type of economic analysis they conducted based on their individual objectives. Objectives for conducting analyses included, for example, identifying economically feasible LID/GI or traditional alternatives, prioritizing LID/GI activities, and alleviating public concerns about the costs of LID/GI approaches. Analysis of the nine case study entities demonstrates that local utilities and governments have found LID/GI practices to be economically preferable to conventional stormwater management practices. These analyses help demonstrate the many and varied kinds of value that can be obtained from LID/GI.

Communities recognize the value of sustainable solutions to stormwater control (e.g., solutions that ensure continued supplies of affordable high-quality drinking water and sustain their community's aquatic resources). Accordingly, rather than always choosing the lowest cost option, communities are making decisions based on the net present value of alternatives over the project life. They are finding that LID/GI offers long-term value. Communities are considering the value of social and environmental benefits in addition to financial costs when assessing their stormwater management approaches. They are enthusiastic about improving their ability to incorporate these benefits into their economic analyses and eager for additional information on best practices to help them in this area. LID/GI provides many environmental and social benefits, and therefore is becoming more accepted as an overall best approach to managing water resources: financially, environmentally, and socially. As LID/GI techniques become more widespread, additional information will become available to improve analysis of these options. However, the studies highlighted throughout this chapter show that there is now: (1) sufficient information available to conduct a range of evaluations, (2) sufficient precedent on study approaches to enable a community to identify a suitable analysis technique for its specific needs, and (3) a substantial demonstration of the need for communities to evaluate a change in the conventional approaches to stormwater management, that can scour our streams; pollute our drinking water reservoirs, rivers, bays, and estuaries; deplete our groundwater resources; and cost our communities their precious and valuable natural resources.

Bibliography

A Triple Bottom Line Assessment of Traditional and Green Infrastructure Options for Controlling CSO Events in Philadelphia's Watersheds, Final Report, Prepared for Howard M. Neukrug, Director, Office of Watersheds, City of Philadelphia Water Department, August 24, 2009.

Alachua County Environmental Protection Department and Public Works Department, Alachua County, FL.

Alachua County Environmental Protection Land Development Regulations page: http://www.alachuacounty.us/government/depts/epd/nr/landregs.aspx.

Alachua County Environmental Protection Land Conservation page: http://www.alachuacounty.us/government/depts/epd/land/.

Alachua County Environmental Protection ACF Site Evaluation Scoring Matrix and criteria: http://www.alachuacounty.us/assets/uploads/images/EPD/Land/ACF_matrix.pdf and http://www.alachuacounty.us/assets/uploads/images/EPD/Land/site_scoring_criteria.pdf.

Beezhold, Michael T., and Donald W. Baker, Rain to Recreation: Making the Case for a Stormwater Capital Recovery Fee, Water Environment Federation WEFTEC Proceedings, 2006.

Bureau of Environmental Services (BES), Portland, Oregon. City of Portland, Bureau of Environmental Services, Sustainable Stormwater Management Program: http://www.portlandonline.com/BES/index.cfm?c = 34598.

Capital Region Watershed District, St. Paul, MN. For more information on CRWD, see: http://www.capitolregionwd.org/.

Charlotte–Mecklenburg Storm Water Services, Mecklenburg County, NC. For more information on projects implemented or planned for implementation in Mecklenburg County, see the Charlotte Mecklenburg Storm Water Services project website: http://www.charmeck.org/Departments/StormWater/Storm+Water+Professionals/Projects.htm.

City of Lenexa Public Works Department, Watershed Division, Lenexa, KS. For more information on Lenexa's Rain to Recreation program, see: www.raintorecreation.org.

City of Los Angeles General Plan (2001): http://cityplanning.lacity.org/.

City of Portland, Bureau of Environmental Services. Cost Benefit Evaluation of Ecoroofs. April 2008. http://www.portlandonline.com/bes/index.cfm?c = 48725&a = 222494.

City of West Union, Iowa and the Iowa Department of Economic Development (IDED).

Department of Environmental Protection (key agency in the development and implementation of the Plan): http://www.nyc.gov/html/dep/html/home/home.shtml.

Fossum, Robert. Volume Reduction BMPs to Address Flooding and Water Quality Problems in the Capitol Region Watershed District, Capitol Region Watershed District, Saint Paul, MN, Proceedings of the Urban Water Management Conference, March 23–23, 2009, Overland Park, KS.

Mayor's Office of Long-term Planning and Sustainability, New York City (NYC).

McDowell Creek Watershed Management Plan Version 4, March 2, 2008.

Open Space Proximity and Land Value: http://www.alachuacounty.us/assets/uploads/images/EPD/Land/Files/Alachua%20Write-up%20Jul%202004.pdf.

Overall PlaNYC: http://www.nyc.gov/html/planyc2030/html/home/home.shtml.

Personal contact information from Mike Beezhold, Watershed Manager, City of Lenexa Department of Public Works.

Personal contact information from Ramesh Buch, Program Manager, Alachua County Forever Land Conservation Program, Alachua County Environmental Protection Department, Gainesville, FL.

Personal contact information from Dave Canaan, Charlotte–Mecklenburg Storm Water Services, NC.

Personal contact information from Mark Doneux, Administrator, Capitol Region Watershed District.

Personal contact information from Angela R. George, Los Angeles River/Ballona Creek, and Richard Gomez, Los Angeles County Department of Public Works.

Personal contact information from Jeff Geerts, Special Project Manager, Iowa Department of Economic Development, Des Moines, IA.

Personal contact information from Aaron Koch, Policy Advisor, Mayor's Office of Long-Term Planning and Sustainability, New York.

Personal contact information from Tom Liptan, Department of Environmental Services, and Terry Miller, Office of Sustainable Development, City of Portland, OR.

Philadelphia's proposed CSO control program: www.phillywatersheds.org. Personal contact information from Howard Neukrug, Director, Office of Watersheds.

Philadelphia Water Department (PWD), Philadelphia, Pennsylvania.

PlaNYC's Sustainable Stormwater Management Plan: http://www.nyc.gov/ html/planyc2030/html/stormwater/stormwater.shtml.

Stormwater BMP Performance Assessment and Cost-Benefit Analysis, January 2010, CRWD.

Sun Valley Watershed, Los Angeles County, CA.

Sun Valley Watershed Management Plan: www.sunvalleywatershed.org.

The University of Florida Program for Resource Efficient Communities has assisted Alachua County in promoting LID and developing land development regulations and enhanced stormwater designs. For more information on this program: http://buildgreen.ufl.edu/.

West Union Green Street Pilot project, see "A Sustainable Vision for West Union: Integrated Green Infrastructure to Achieve A Renaissance of West Union's Downtown District and Neighborhoods" at: http://www.westunion.com/ uploads/PDF_File_67184288.pdf.

6

The Property-Price Effects of Abating Nutrient Pollutants in Urban Housing Markets

Patrick Walsh, J. Walter Milon, and David Scrogin

CONTENTS

Introduction

The state of Florida has experienced rapid population growth over the past 50 years. The most significant increase has occurred since 2000, with the population rising by more than 16%. This growth has been accompanied by increased residential and commercial development and investment in public infrastructure. As has been mentioned in several previous chapters, when landscape is transformed and the levels of impervious surface increase, so too do the levels of nutrient pollutants in local waters due to stormwater runoff. Florida has a long history of stormwater regulation, but nutrient loading has emerged as the primary threat to inland waters, and most notably so in the central Florida corridor (OCEPD, 2007). Several programs have recently been introduced at the state and local levels to mitigate the negative environmental impacts of stormwater runoff. Among these are programs that give property owners incentives to abate runoff from their residences and allow neighborhoods to vote on municipal service taxes to fund water quality management at nearby lakes.

127

In addition to state and local initiatives, a forthcoming federal rule has changed the regulatory environment in Florida, with the EPA proposing to replace the long-standing narrative nutrient criteria with numeric nutrient criteria (Federal Register, 2010). All water bodies are to be assigned strict limits for the content of several nutrients linked to stormwater runoff, including nitrogen, phosphorous, and chlorophyll *a*. This regulation has the potential to improve water quality throughout the state and is expected to require substantial control of nonpoint source pollution (FDEP, 2008). As the state is expected to incur significant costs in achieving the proposed standards, it is important to recognize the variety of benefits that can result from stormwater abatement and reductions in nutrient pollutants. This chapter investigates one dimension of these benefits: property price changes in urban housing markets.

Hedonic property price models are estimated with a large sample of property sales occurring in Orange County, Florida over the period 1996–2004. Of interest is measuring the relationship between property prices and several continuous indicators of water quality in local lakes, including the trophic state index (TSI), a composite of three nutrient pollutants resulting from stormwater runoff (for details on calculation, see pp. 86–87 of Florida's 1996 305(b) report). Given that properties located throughout the housing market contribute to runoff deposited in local lakes and the costs of achieving the standards will eventually be shared by taxpayers, our analysis includes both waterfront and nonwaterfront property sales. Results indicate the property price effects of stormwater pollution abatement (i) differ significantly between lakefront and nonlakefront properties, (ii) diminish with distance from the waterfront, and are (iii) an increasing function of the size of the water body. The hedonic valuations of water quality and stormwater management in Florida are reviewed in the next section. Following this, the hedonic property price models are specified and the dataset and summary statistics are discussed. Estimation results and the marginal implicit prices of reduced nutrient levels are discussed and a conclusion is reached.

Literature Review and Background

Hedonic price analyses have been conducted in various settings, ranging from pricing automobiles (Atkinson and Halvorsen, 1984) to the wages paid in labor markets (Garen, 1988). In the nonmarket valuation literature, hedonic models have been used to measure the property price effects of environmental amenities and disamenities, including proximity to hazardous waste sites (Kolhase, 1991), improvements in air quality

(Smith and Huang, 1995; Boyle and Kiel, 2001), and access to open space (Cho et al., 2009). The methodology was pioneered by Rosen (1974), who describes equilibrium between profit-maximizing producers and utility-maximizing consumers. A useful feature of the model for analysis of public programs is that under certain conditions the mean implicit price of (or marginal willingness to pay for) individual attributes of the good may be estimated.

If water quality affects the sales prices of surrounding properties, then hedonic models may be used to estimate the impact of pollution abatement efforts on property prices. Many air quality studies have been published, however, the water quality valuation literature remains sparse. This may be because the health risks of exposure to waterborne pollutants may be small relative to those of airborne pollutants and because sufficient variation in water quality over time or water bodies is needed to estimate values with defensible statistical significance.

Early water quality studies used surveys of homeowners, water biologists, or government officials to construct proxies of water quality, including David (1968), Epp and Al-Ani (1979), and Young (1984). As water sampling became more prevalent and the measurement techniques more consistent, studies of housing markets throughout the country began to employ both physical indicators, including concentrations of dissolved oxygen and nitrogen (Brashares, 1985), fecal coliform (Leggett and Bockstael, 2000), suspended solids (Poor, Pessagno, and Paul, 2007), and measures of water clarity or transparency, most notably secchi depth (Michael, Boyle, and Bouchard, 1996, 2000; Boyle, Poor, and Taylor, 1999; Poor et al., 2001; Gibbs, Halstead, and Boyle, 2002; Walsh, Milon, and Scrogin, 2010). Several studies have directly examined a stormwater's impact on property prices. Loomis and Streiner (1995) and Braden and Johnston (2004) investigated the benefits of flood reduction resulting from stormwater management. Bin, Landry, and Meyer (2009) focused on the impact of riparian buffers adjacent to streams on property prices. And Sander, Polasky, and Haight (2010) found that reduced stormwater flows resulting from increased residential tree cover could significantly affect property prices.

In Florida, nutrient pollutants from stormwater runoff from residential and commercial development have grown to be the greatest threat to the quality of public waters (FDEP, 1996, 2006, 2008). The state has a long history of managing and regulating stormwater flows and was first in the nation to implement a stormwater permitting program for all development. The Florida Stormwater Rule took effect in 1982 and required all development except for single-family residential projects to obtain stormwater permits and abate discharges. Figure 6.1 (FDEP, 2008) illustrates the impact of the rule on phosphorous levels and reveals a recent upswing since 2000. This period has experienced considerable population growth

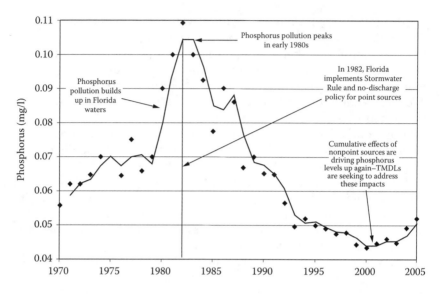

FIGURE 6.1
Average phosphorous in 3,300 Florida surface water bodies. (From FDEP. (2008). Integrated Water Quality Assessment for Florida: 2008 305b Report and 303 d List Update. Tallahassee: Florida Department of Environmental Protection.)

with increased residential development throughout the state. New programs have since been initiated to mitigate the negative environmental impacts, inasmuch as single-family residential development was exempt from the rule.

One of the programs administered in Orange County is investigated in this chapter. This particular program dates to the 1970s and allows neighborhoods to vote on imposing a property tax to fund projects at specific lakes, including stormwater management and aquatic plant control. Support by 66% of homeowners leads to adoption of either a municipal service tax unit (MSTU) or a municipal service benefit unit (MSBU) within the neighborhood. The MSTU takes the form of a millage on residential property taxes so assessed property values dictate the tax payment, whereas the MSBU is a flat tax paid by all residential property owners within the neighborhood (about $150 per household). There are 21 lakes managed through the MSTU program and 10 lakes managed under the MSBU program. As well as the MSTU/MSBU program, Orange County instituted the Financial Incentives Program in 2004 as part of Florida's Clean Lakes Initiative. This program provides homeowners with a rebate of up to $1,000 dollars for installing a stormwater berm and swale system, restoring lakeshore vegetation and littoral zones, and removing upland invasive tree species. Public participation exhausted the program's $150,000 budget in 2008.

In the next section, we examine the benefits of public efforts to reduce nutrient levels in Orange County lakes as reflected through residential property prices. The nutrient indicators include total nitrogen (TN), total phosphorous (TP), and chlorophyll *a* (CH). Elevated levels of these indicators can lead to reduced water clarity and foul odor, reduced levels of oxygen, and increased levels of bacteria, and ultimately undermine the health of the aquatic and surrounding terrestrial ecosystems (USEPA, 2003). These pollutants appear in the Florida numeric nutrient criteria that were proposed by the EPA in 2009.[*] The criteria assign numeric thresholds to lakes, streams, springs, and canals for nutrient content, effectively replacing the state's long-standing *narrative* nutrient criteria. TN, TP, and CH are the inputs of the fourth water quality variable used in this chapter: the trophic state index (TSI). The TSI was originally developed by Carlson (1977) to be a comprehensive indicator of lake health.[†] Similar indicators have appeared in other water quality valuation studies (Pendleton, Martin, and Webster, 2001; Massey, Newbold, and Gentner, 2006) and have a long history in the conservation biology literature (Ribaudo et al., 2001), including studies of tropical forestation (Sheil, Nasi, and Johnson, 2004), marine ecosystem quality (Muxika et al., 2005; Blocksom and Johnson, 2009), and biodiversity (Feest, Aldred, and Jedamzik, 2010).

Model Specification

The hedonic property price model specifies the selling price of a property as a stochastic function of structural characteristics of the property, environmental amenities, and other locational attributes within the housing market. In specifying the model and obtaining the implicit water quality value, households are assumed to be freely mobile within the market to choose housing bundles that maximize utility over the attributes (Kuminoff, 2009). To allow the implicit price of water quality to vary between properties and between water bodies, water quality enters the model through multiple interaction variables. The linear-in-parameters, linear-in-variables specification of the model is:

---

[*] http://www.epa.gov/waterscience/standards/rules/florida/
[†] The TSI used here is calibrated to local Florida conditions such as tannin-induced "dark water," and accounts for the nutrient-limiting concept. A TSI value between 0–59, 60–69, or 70–100 indicates, respectively, the biological health of the water body is good, fair, or poor.

$$P = \beta_0 + \beta_1 Waterfront + \beta_2 WQ + \beta_3 Waterfront * WQ + \beta_4 Distance + \beta_5 Distance * WQ$$
$$+\beta_6 Area + \beta_7 Area * WQ + \beta_8 MSTU + \gamma'S + \phi'N + \psi'l + \delta't + \varepsilon$$

(6.1)

where P references the sales price; *Lakefront* is a dummy variable distinguishing lakefront from nonlakefront properties; *WQ* is a continuous measure of water quality in the nearest lake at the time of sale; *Distance* is a measure of the property proximity to the waterfront; *Area* is the surface water coverage of the nearest lake; *MSTU* indicates whether the home pays an MSTU/MSBU fee; S is a vector of structural attributes; N is a vector of location and neighborhood attributes; t is a vector of time dummy variables; l is a vector of lake dummy variables; and ε is random error.

The model is estimated separately for each water quality variable due to a high degree of correlation between TN, TP, and CH in the sample, highlighting another potential advantage of using comprehensive trophic state indices for hedonic water quality valuation. Water quality enters (1) as its own variable and through interactions with several other variables. The interaction with the lakefront dummy reflects the hypothesis that water quality affects waterfront property prices differently from nonwaterfront property prices due to their unique location within the landscape. The interaction with the distance variable reflects the hypothesis that the effect of water quality on property price depends on the property's proximity to the nearest lake. Lastly, the interaction with the area variable reflects the hypothesis that the effect of water quality on property price depends on the size of the lake. A similar specification of the model that includes these three interaction variables was investigated by Walsh et al. (2011).

The lake dummy variables control for lake-specific attributes that do not vary over time, such as private or public access or the existence of a boat launch. The main tradeoff with these variables is that the lake area variable cannot be used alone in the regression (although it still appears as an interaction), because it does not change over time. However, as long as the lake area does not change significantly over the study area, its impact should be captured with the lake dummies. The one caveat is there is the risk that the area interaction with water quality may overstate the impact of the lake area on the value of water quality.

Two final modeling issues are potential spatial autocorrelation and functional form specification. Spatial autocorrelation can be induced by unobserved neighborhood and landscape attributes common to property sales located within close proximity to one another. In the presence of spatial dependence, the assumption of independent errors does not hold and the parameter estimates are inconsistent (Anselin, 1999). This concern has been raised in several hedonic property studies, with spatial lag and spatial error econometric models estimated to test and correct for spatial dependence in the errors.[*] The findings are mixed, with some studies reporting improvements in fit (Gelfand et al., 2004; Case et al., 2004) and others reporting results that are robust to the error specification (Leggett and Bockstael, 2000; Kim, Phipps, and Anselin, 2003; Mueller and Loomis, 2008). For completeness, estimation is performed under the null of zero spatial correlation and with the spatial lag specification. Inasmuch as it has been shown that the specification of the spatial weights matrix (SWM) can influence regression results (Bell and Bockstael, 2000), two variations of SWM are used in the present chapter; one based on inverse distance and the other based on the 15 nearest neighbors.[†] Maximum likelihood is used to estimate the spatial lag models, as explained in LeSage (1999).

The functional form of the model must be specified. Although there is no theoretical basis for the functional form, Cropper, Deck, and McConnell (1988) suggest that semi-log and double-log models are superior to more complicated models in the face of misspecification and proxy variables. Both regular Box–Cox tests and spatially robust Box–Cox tests using double-length regression tests (Le and Li (2008) and Le (2009)) favored the double-log specification appearing in expression (6.1). Furthermore, both hedonic (Michael, Boyle, and Bouchard, 2000) and limnology (Hoyer et al., 2002) sources recommend transforming the water quality variables into natural logarithms because marginal changes may be more recognizable at low levels.

Data

The data include 146 Orange County lakes that vary in size from approximately 1 acre to 1,800 acres.[‡] Table 6.1 contains a summary of the sizes of the included lakes. All the water quality indicators were obtained from EPA's STORET database, with missing observations obtained from three municipalities. Table 6.2 contains summary statistics on the water quality

---

[*] The spatial lag model is written $Y = \rho WY + X\beta + u$, where Y is the dependent variable, W is a spatial weights matrix, X is a vector of regressors, β and ρ are parameters to be estimated, and u is an i.i.d. normal random error. The spatial error model appears as $Y = X\beta + \gamma$, where $\gamma = \lambda W\gamma + u$, and β and λ are parameters. Estimates of ρ and λ may be used to test for spatial dependence in the errors. See Anselin (1988) and LeSage (1999) for technical details on spatial econometric models.

[†] The inverse distance SWM is created with appraisal practices in mind; homes sold within a distance of a half mile and within a time period of six months previous and three months forward are considered neighbors. If property j is a neighbor to property i, $W_{ij} = 1/d_{ij}$, where d_{ij} is the distance between the two homes. The nearest neighbor specification uses $W_{ij} = 1$ if property j is one of the 15 nearest properties to home i. Both types of SWM are row-standardized.

[‡] Only named lakes are used in the analysis.

TABLE 6.1

Lake Size Description

Size (Acres)	Number of Lakes	Average Size
0–50	82	16.72
50–100	23	74.16
100–300	26	162.35
>300	15	818.10
Total	146	284.74

variables and illustrates significant variation in each variable over the nine-year sample period.

The property data were obtained from the Orange County Property Appraiser (OCPA). These data contain all single-family property sales during the time period 1996–2004. All single-family home sales within 1,000 meters of the 146 lakes are used in the analysis, yielding 54,716 total property sales.* The property data contain added home attribute data, including the number of bedrooms and bathrooms, size of the parcel, size of the home, home age, the existence of a pool, and if the home is classified as lakefront, canal-front, or golf-front. The OCPA provided latitude and longitude coordinates of each home, which were geo-coded on GIS maps to calculate the distance to the central business district (CBD: downtown Orlando) and distance to the nearest lake.† Through the OCPA data, it is also possible to identify which homes pay an MSTU/MSBU fee. Thirteen percent of the properties analyzed in the present chapter pay an MSTU or MSBU fee. Parcels were mapped to census tracts to obtain neighborhood variables such as income, race, and age demographics. Summary statistics appear in Table 6.3.

Estimation Results

The estimated water quality coefficients appear in Table 6.4.‡ For each water quality variable, three models are estimated: ordinary least squares (OLS), a spatial regression with the nearest neighbor SWM (NN15), and a spatial regression with an inverse-distance SMW (DT). The models provide nearly identical fits to the data as reflected by the R^2 statistics, indicating the models are robust to the error specification. A likelihood ratio test

* Properties that sold for less than $30,000 were excluded from the dataset.
† There was concern that the distance to other lakes was also relevant. A lake count variable was used in preliminary regressions, which expressed the number of lakes within 1,000 meters from each parcel. This variable was insignificant in all specifications.
‡ The full set of estimation results is available in an appendix upon request.

TABLE 6.2

Water Quality Summary Statistics

	TSI	TN	TP	CH
1996–1998				
Mean	44.756	0.788	0.035	11.484
St. Dev	12.622	0.287	0.034	11.044
Min	16.470	0.220	0.003	0.681
Max	77.175	2.281	0.299	72.431
1999–2001				
Mean	42.488	0.845	0.034	9.892
St. Dev	14.529	0.334	0.034	11.684
Min	11.190	0.154	0.005	0.350
Max	87.495	2.565	0.350	69.331
2002–2004				
Mean	46.351	0.839	0.038	11.411
St. Dev	11.978	0.360	0.031	10.583
Min	11.580	0.176	0.004	0.500
Max	78.070	2.450	0.350	67.058

(LeSage, 1999) rejects the hypothesis of no spatial dependence at the 99% level.[*] The spatial lag coefficient (ρ) is small but significant in each model.

In all models, the waterfront and water quality interaction term is significant and negative, indicating the mean marginal value of water quality is greater for waterfront homes. The distance interaction term is positive and significant in all specifications, indicating the marginal value of water quality decreases as a property's distance to the nearest lake increases. The hypothesis about the interactive effects of water quality and lake size is supported for TN, CH, and TSI but is rejected for TP. It is of the correct sign, but insignificant, in the TP models. Overall, the pollutants targeted by the numeric nutrient criteria proposed by the EPA are significantly related to property prices. Finally, in all specifications the MSTU/MSBU coefficient is positive and significant, suggesting that neighborhoods participating in the program realize price effects from water quality improvements beyond those of nonparticipants.

---

[*] Although robust Lagrange multiplier tests are normally used to select between the spatial error and lag models, the size of the present dataset is prohibitively large for that procedure. Using the same data, Walsh (2009) uses out-of-sample prediction to compare the two models, gauged by six categories of prediction error from Case et al. (2004). The spatial lag model outperforms the spatial error model in five of the six categories. Furthermore, there were only minor differences in the magnitudes of the coefficients between the spatial error and spatial lag models.

TABLE 6.3

Descriptive Statistics (N = 54,712)

Variable	Units	Mean	Std. Dev.	Min	Max
Sales Price	2002 Dollars	206,891.00	211,640.90	30,048	3,986,184
Heated Area	Square Feet	1,984.096	969.632	506	23,728
Area of Parcel	Square Meters	1,122.693	1,205.603	53.510	98,582.010
% Pay MSTU/ MSBU		13.08			
% Waterfront		2.7			
Number of Bedrooms		3.304	0.844	1	16
Home Age	Years	18.759	14.992	0.5	99
% With Pool		20.7			
Distance to Nearest Lake	Meters	455.525	273.013	2	1,130.420
Area of Nearest Lake	Acres	284.736	453.150	0.997	1,853.161
Distance to CBD	Meters	9,253.748	5,149.971	506.290	28,589.640
% in airport noise zone		15.7			
Latitude Coordinate	Degrees	653,265.700	7,930.728	635,754.200	683,568.200
Longitude Coordinate	Degrees	505,663.300	6,285.382	487,273.600	517,649.500
% Canal-front		0.3			
% Golf-front		0.1			
% of Population White		78.2			
% of Population Black		12.3			
% of Population > 65		11.2			
Median Household Income	2002 Dollars	56,578.670	23,993.800	9,641	174,169
% of sales in 1996		9.2			
% of sales in 1997		10.9			
% of sales in 1998		12.2			
% of sales in 1999		13.4			
% of sales in 2000		11.4			
% of sales in 2001		9.9			
% of sales in 2002		9.9			
% of sales in 2003		11.0			
% of sales in 2004		12.2			

TABLE 6.4

Selected Hedonic Estimation Results

		WQ *WF	Dist *WQ	Area *WQ	Pay MSTU	ρ	R²
	WQ						
TN							
OLS	−0.083** (0.042)	−0.125* (0.025)	0.032* (0.004)	−0.009* (0.003)	0.047* (0.005)	—	0.8938
NN15	−0.094** (0.041)	−0.124* (0.019)	0.032* (0.004)	−0.008* (0.003)	0.046* (0.005)	−0.003* (0.000)	0.8943
DT	−0.081** (0.041)	−0.127* (0.019)	0.032* (0.004)	−0.009* (0.003)	0.047* (0.005)	0.001* (0.000)	0.8937
TP							
OLS	−0.088* (0.032)	−0.124* (0.012)	0.019* (0.002)	−0.002 (0.002)	0.034* (0.005)	—	0.8943
NN15	−0.102* (0.029)	−0.123* (0.009)	0.018* (0.002)	−0.001 (0.002)	0.034* (0.005)	0.003* (0.000)	0.8948
DT	−0.090* (0.029)	−0.124* (0.009)	0.019* (0.002)	−0.002 (0.002)	0.034* (0.005)	0.001* (0.000)	0.8943
CH							
OLS	−0.014 (0.015)	−0.064* (0.008)	0.010* (0.001)	−0.004* (0.001)	0.040* (0.005)	—	0.894
NN15	−0.015 (0.014)	−0.064* (0.005)	0.010* (0.001)	0.004* (0.001)	0.039* (0.005)	−0.003* (0.000)	0.8946
DT	−0.015 (0.014)	−0.015* (0.014)	0.010* (0.001)	−0.004* (0.001)	0.040* (0.005)	0.001* (0.000)	0.8941
TSI							
OLS	−0.164* (0.051)	−0.168* (0.025)	0.045* (0.005)	−0.008* (0.003)	0.032* (0.005)	—	0.8943
NN15	−0.176* (0.048)	−0.167* (0.016)	0.046* (0.004)	−0.007** (0.003)	0.032* (0.005)	−0.003* (0.000)	0.8948
DT	−0.168* (0.048)	−0.169* (0.016)	0.046* (0.004)	−0.008* (0.003)	0.032* (0.005)	0.001* (0.000)	0.8943

Note: N = 54,712. *, **, and *** denote significance at the 1%, 5%, and 10% levels, respectively. Standard errors appear in parentheses.

The water quality variables are all expressed in different units, thus the percentage changes implied by the various coefficients in Table 6.4 are not easily compared. The marginal effect, or implicit price, of a change in each of the water quality indicators is more useful. From the spatial lag specification of the model (6.1), the implicit value of water quality may be written:

$$\frac{\partial P}{\partial WQ} = \left(\frac{1}{1-\rho}\right)\left(\frac{P}{WQ}\right)\left(\beta_2 + \beta_3 * WF + \beta_5 * \ln(Distance) + \beta_7 * \ln(Area)\right) \quad (6.2)$$

TABLE 6.5

Implicit Price of a 17% Change in Indicator

		Mean Waterfront Implicit Price	Mean Nonwaterfront Implicit Price
TN			
	OLS	8,076.10 (857.24)	378.64 (167.31)
	NN15	7,870.13 (612.84)	363.96 (157.68)
	DT	7,650.60 (616.44)	242.59 (156.80)
TP			
	OLS	5,812.44 (434.55)	—
	NN15	5,736.77 (318.74)	—
	DT	5,883.09 (320.57)	—
CH			
	OLS	3,217.50 (258.68)	28.12 (59.19)
	NN15	3,239.11 (177.21)	23.51 (51.75)
	DT	3,261.50 (178.25)	9.74 (52.06)
TSI			
	OLS	9,673.93 (810.33)	—
	NN15	9,573.05 (515.81)	—
	DT	9,729.90 (518.86)	—

Note: Standard errors appear in parentheses.

Estimates of the implicit value of marginal changes in TN, TP, CH, and TSI may be obtained by substituting the coefficient estimates reported in Table 6.4 and the sample means of the respective independent variables into (6.2). Given the dependent variable is in natural logarithms, an estimate of $E(P|X)$ obtained with the Duan (1983) smearing estimator is used in place of the sample mean of property price. Table 6.5 reports the mean implicit prices of water quality for lakefront and nonlakefront properties obtained from the spatial and nonspatial specifications of the model. To facilitate comparison across indicators and with past Secchi disk measurement (SDM) literature (Michael et al., 2000; Poor et al., 2001; Gibbs et al., 2002; Krysel et al., 2003), implicit prices are expressed as a 17% change.* Because the Florida numeric nutrient criteria imply different changes for each affected lake, the current approach should provide a tool that can be adapted to individual conditions.

---

* For comparison to other hedonic papers, a 1-ft change in SDM corresponds to a 17% change in the SDM level of the average lake in the sample.

Table 6.5 indicates that the spatial models do not cause large differences in the magnitudes of the implicit prices, although the standard errors of the spatial waterfront implicit prices are notably lower than those of the nonspatial specification. TSI has the highest waterfront implicit price at approximately $9,600, equivalent to 5% of the average home price (2% of the average waterfront price of $452,646). Although only slightly larger than the implicit price of TN, this is roughly three times larger than the CH waterfront implicit price ($3,200). The effect of waterfront on the implicit prices is clear from the table, where TN has the largest nonwaterfront prices. At the average nonwaterfront lake distance of 456 meters, the implicit prices of the other indicators are essentially zero. However, the nonwaterfront implicit prices for homes closer to the lake are considerably larger. To better illustrate the impact of distance on the implicit prices, Figure 6.2 contains a graph of the implicit price gradients of each indicator. At 100 meters from the lake, TSI has an implicit price of $1,800, with TN slightly less. TN has the farthest extent, with positive implicit prices beyond 600 meters from the lake. TP has the shortest extent, going to zero at approximately 350 meters away. Nonetheless, all the water quality variables yield positive nonwaterfront prices for homes several hundred meters around a lake.

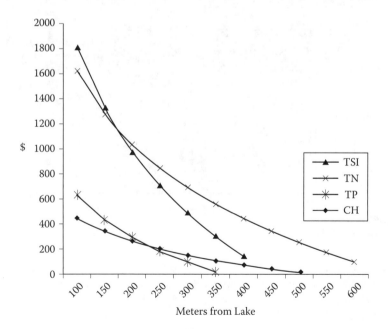

FIGURE 6.2
Distance, water quality, and property prices.

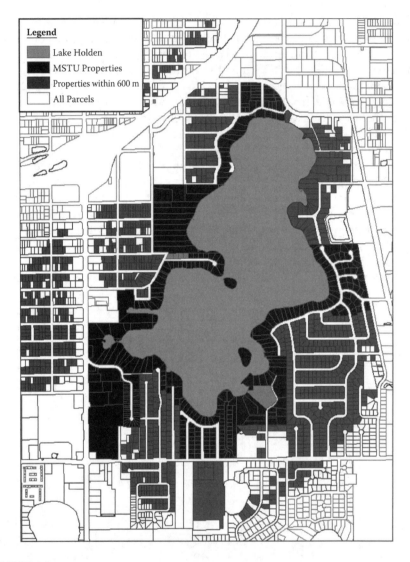

FIGURE 6.3
Lake Holden MSTU.

An interesting application is a comparison of MSTU boundaries to the extent of benefits implied by the implicit prices. Figure 6.3 contains a map of the parcels surrounding Lake Holden, with those that pay the MSTU fee highlighted in black. Single family homes (with a sales price above $30,000 and below $400,000) within 600 meters of the lake are colored gray to illustrate the potential implicit price gradient, and other parcels are white. In this map, there appears to be a positive externality from the MSTU; water quality increases from MSTU projects can affect a substantial number of added homes not enrolled in the MSTU. Homes that pay the MSTU see a property price increase from the water quality improvement, as well as the 3% increase from simply being in the MSTU. This policy appears to be efficient, inasmuch as the annual fee paid by homes is a fraction of the property price appreciation that results.

Discussion and Conclusions

The State of Florida has had an expansive program of water quality management over the last half century. Both point and nonpoint sources of pollution have been regulated, with the latter increasing in importance since the 1982 Stormwater Rule. However, of those water bodies evaluated by the FDEP, 28% of river and stream miles, 25% of lake acres, and 59% of estuary square miles are classified as having poor water quality (FDEP, 2008). Furthermore, 24% of assessed waters are degrading over time. Urban stormwater is a persistent threat to Florida waters, contributing nutrients to water bodies through a diverse array of sources. The EPA's proposed numeric nutrient criteria are designed to halt the progress of nutrient pollution by imposing standards on all water bodies, replacing the previous narrative criteria that required time and resource-intensive waterbody-specific analysis. The present chapter analyzed potential benefits to improvements in the targets of the new criteria through a hedonic property analysis.

The results presented here indicate that property prices are significantly affected by the pollutants targeted by the numeric nutrient criteria. Four indicators of nutrient pollution were employed, with a new compound variable, TSI, introduced to the hedonic literature. The marginal values of these pollutants were found to be considerably larger for waterfront homes and to diminish with distance to the lake. Furthermore, the marginal value of water quality increased with lake area in three of the four quality variables. Results are supportive of the compound TSI variable, because it could incorporate changes in TN, TP, and CH, while accounting for the nutrient limiting concept. The TSI represents a useful indicator for future benefits analysis in Florida and other areas, inasmuch as it can be calibrated to account for local conditions. For instance, Ohio uses a TSI that has been calibrated using SDM instead of chlorophyll a.

The estimated implicit prices illustrate incentives for homeowners to contribute to stormwater improvement activities, made possible by Orange County policies such as the MSTU/MSBU program. Results indicate that homes in the program receive both a direct and indirect benefit from participating. For the direct benefit, property prices increase an average of 3–5% for participating in the MSTU/MSBU program. Participation may signal to potential home buyers that the community is more environmentally conscious than others, or it may be seen as an assurance that lake water quality will be maintained in the future. The indirect benefit occurs if the MSTU/MSBU program is successful at improving water quality, which will increase property values further. Moreover, this indirect effect is a positive externality of the program, since it will benefit other nearby homes that are not enrolled. As the returns to traditional stormwater regulation decrease over time, policies such as those instituted in Orange County may represent complementary methods of combating stormwater runoff.

Acknowledgments

We are grateful to Jesse Abelson for assistance in constructing the water quality dataset and Mark Schneider for the GIS programming and to Shelby Gerking, Mark Dickie, and seminar participants at the CU Environmental and Resource Economics workshop and the Northeastern Agricultural and Resource Economics Association Annual Conference. The views expressed in this chapter and any errors should be attributed solely to the authors.

References

Anselin, L. (1988). *Spatial Econometrics: Methods and Models*. Norwell, MA: Kluwer Academic.

Anselin, L. (1999). Spatial Econometrics. Working paper. University of Texas at Dallas. 1–31.

Atkinson, S.E., and R. Halvorsen. (1984). A new hedonic technique for estimating attribute demand: An application to the demand for automobile fuel efficiency. *Rev.Econ. Statist.*, 66(3): 417–426.

Bell, K.P. and N.E. Bockstael. (2000). Applying the generalized-moments estimation approach to spatial problems involving microlevel data. *Rev. Econ. Statist.*, 82(1): 72–82.

Bin, O., C.E. Landry, and G.F. Meyer. (2009). Riparian buffers and hedonic prices: A quasi-experimental analysis of residential property values in the Neuse River basin. *Amer. J. Agric. Econ.*, 91(4): 1067–1079.

Blocksom, K.A., and B.R. Johnson. (2009). Development of a regional microinvertebrate index for large river bioassessment. *Ecol. Indicat.*, 9(2): 313–328.

Boyle, K.J., P.J. Poor, and L.O. Taylor. (1999). Estimating the demand for protecting freshwater lakes from eutrophication. *Amer. J. Agric. Econ.*, 81(5): 1118–1122.

Boyle, M.A., and K.A. Kiel. (2001). A survey of house price hedonic studies of the impact of environmental externalities. *J. Real Estate Lit.*, 9(2): 117–144.

Braden, J.B., and D.M. Johnston. (2004). Downstream economic benefits from storm-water management. *J. Water Res. Plan. Manage.*, 130(6): 498–505.

Brashares, E. (1985). Estimating the Instream Value of Lake Water Quality in Southeast Michigan. Ph.D. dissertation, Department of Economics, University of Michigan.

Carlson, R.E. (1977). A trophic state index for lakes. *Limnol. Oceanography*, 22(2): 361–369.

Case, B., J. Clapp, R. Dubin, and M. Rodriguez. (2004). Modeling spatial and temporal house price patterns: A comparison of four models. *J. Real Estate Fin. Econ.*, 29(2): 167–191.

Cho, S.-H., C.D. Clark, W.M. Park, and S.G. Kim. (2009). Spatial and temporal variation in the housing market values of lot size and open space. *Land Econ.*, 85(1): 51–73.

Cropper, M.L., L.B. Deck, and K.E. McConnell. (1988). On the choice of functional form for hedonic price functions. *Rev. Econ. Statist.*, 70: 668–675.

David, E.L. (1968). Lakeshore property values: A guide to public investment in recreation.*Water Resources Res.*, 4(4): 697–707.

Duan, N. (1983). Smearing estimate: A nonparametric retransformation method. *J. Amer. Statist. Assoc.*, 78: 605–610.

Epp, D.J., and K.S. Al-Ani. (1979). The effect of water quality on rural nonfarm residential property values. *Amer. J. Agric. Econ.*, 61(3): 529–534.

FDEP. (1996). 1996 Water-Quality Assessment for the State of Florida: Section 305 b Main Report. Tallahassee: Florida Department of Environmental Protection.

FDEP. (2006). Integrated Water Quality Assessment for Florida: 2006 305 (b) Report and 303 (d) List Update. Tallahassee: Florida Department of Environmental Protection.

FDEP. (2008). Integrated Water Quality Assessment for Florida: 2008 305 b Report and 303 d List Update. Tallahassee: Florida Department of Environmental Protection.

Federal Register. (2010). Vol. 75, No. 233, December 6, 2010, FRL-9228-7.

Feest, A., T.D. Aldred, and K. Jedamzik. (2010). Biodiversity quality: A paradigm for biodiversity. *Ecol. Indicat.*, 10(6): 1077–1082.

Garen, J. (1988). Compensating wage differentials and the endogeneity of job riskiness. *Rev. Econ. Statist.*, 70(1): 9–16.

Gelfand, A.E., M.D. Ecker, J.R. Knight, and C.F. Sirmans. (2004). The dynamics of location in home price. *J. Real Estate Fin. Econ.*, 29: 149–166.

Gibbs, J.P., J.M. Halstead, and K.J. Boyle. (2002). An hedonic analysis of the effects of lake water clarity on New Hampshire lakefront properties. *Agric. Resource Econ. Rev.*, 31(1): 39–46.

Hoyer, M.V., T.K. Frazer, S.K. Notestein, and D.E. Canfield, Jr. (2002). Nutrient, chlorophyll, and water clarity relationships in Florida's nearshore coastal waters with comparisons to freshwater lakes. *Canad. J. Fish. Aquat. Sci.*, 59: 1024–1031.

Kim, C.W., T.T. Phipps, and L. Anselin. (2003). Measuring the benefits of air quality improvement: A spatial hedonic approach. *J. Environ. Econ. Manage.*, 45:
24–39.

Kolhase, J. E. (1991). The impact of toxic waste sites on housing values. *J. Urban.
Econ.*, 30(1): 1–26.

Krysel, C., E.M. Boyer, C. Parson, and P. Welle. (2003). Lakeshore Property
Values and Water Quality: Evidence from Property Sales in the Mississippi
Headwaters Region. Legislative Commission on Minnesota Resources.
Bemidji, Mississippi Headwaters Board and Bemidji State University: 59.

Kuminoff, N.V. (2009). Decomposing the structural identification of non-market
values. *J. Environ. Econ. Manage.*, 57: 123–139.

Le, C.Q. (2009). Three Essays in Spatial Econometrics and Labor Economics, PhD
dissertation, Department of Economics, Kansas State University.

Le, C.Q., and D. Li. (2008). Double length regression tests for testing functional
form and spatial error dependence. *Econ. Lett.*, 101: 253–257.

Leggett, C.G., and N.E. Bockstael. (2000). Evidence of the effects of water quality
on residential land prices. *J. Environ. Econ. Manage.*, 39: 121–144.

LeSage, J.P. (1999). The theory and practice of spatial econometrics. Documentation
for Spatial Econometrics Toolbox for MATLAB. Ohio: University of Toledo.

Loomis, J.B., and C.F. Streiner. (1995). Estimating the benefits of urban steam restoration using the hedonic price method. *Rivers*, 5(4): 267–278.

Massey, D.M., S.C. Newbold, and B. Gentner. (2006). Valuing water quality changes
using a bioeconomic model of a coastal recreation fishery. *J. Environ. Econ.
Manage.*, 52: 482–500.

Michael, H.J., K.J. Boyle, and R. Bouchard. (1996). Water Quality Affects Property
Prices: a case Study of Selected Maine Lakes. Miscellaneous Report 398,
Maine Agricultural and Forest Experiment Station, University of Maine.

Michael, H.J., K.J. Boyle, and R. Bouchard. (2000). Does the measurement of environmental quality affect implicit prices estimated from hedonic models?
Land Econ., 76(2): 283–298.

Mueller, J.M., and J.B. Loomis. (2008). Spatial dependence in hedonic property models:
Do different corrections for spatial dependence result in economically significant
differences in estimated implicit prices? *J. Agric. Resource Econ.*, 33: 212–231.

Muxika, I., Á. Borja, and W. Bonne. (2005). The suitability of the marine biotic
index (AMBI) to new impact sources along European coasts. *Ecol. Indicat.*,
5(1): 19–31.

OCEPD. (2007). 2007 Water Quality Summary Report. Orange County
Environmental Protection Department. Orlando, FL.

Pendleton, L., N. Martin, and D.G. Webster. (2001). Public perceptions of environmental quality: A survey study of beach use and perceptions in Los Angeles
County. *Marine Pollut. Bull.*, 42(11): 1155–1160.

Poor, P.J., K.J. Boyle, L.O. Taylor, and R. Bouchard. (2001). Objective versus subjec-
tive measures of water clarity in hedonic property value models. *Land Econ.*,
77(4): 482–493.

Poor, P J., K.L. Pessagno, and R.W. Paul. (2007). Exploring the hedonic value of
ambient water quality: A local watershed-based study. *Ecol. Econ.*, 60(4):
797–806.

Ribaudo, M.O., D.L. Hoag, M.E. Smith, and R. Heimlich. (2001). Environmental indices and the politics of the conservation reserve program. *Ecol. Indicat.,* 1(1): 11–20.

Rosen, S. (1974). Hedonic prices and implicit markets: Product differentiation in pure competition. *J. Polit. Econ.,* 82(1): 34–55.

Sander, H., S. Polasky, and R.G. Haight. (2010). The value of urban tree cover: A hedonic property price model in Ramsey and Dakota Counties, Minnesota, USA., *Ecol. Econ.,* 69: 1646–1656.

Sheil, D., R. Nasi, and B. Johnson. (2004). Ecological criteria and indicators for tropical forest landscapes: Challenges in the search for progress. *Ecol. Soc.,* 9(1): 1–7.

Smith, V.K., and J.-C. Huang. (1995). Can markets value air quality? A meta-analysis of hedonic property value models. *J. Polit. Econ.,* 103(1): 209–227.

USEPA. (2003). National Management Measures for the Control of Nonpoint Pollution from Agriculture. Office of Water. Washington DC: US Environmental Protection Agency.

Walsh, P. (2009). Hedonic Property Value Modeling of Water Quality, Lake Proximity, and Spatial Dependence in Central Florida. PhD dissertation. Department of Economics, University of Central Florida.

Walsh, P., J. W. Milon, and D. Scrogin. (2011). The Spatial Extent of Water Quality Benefits in Urban Housing Markets, *Land Economics,* forthcoming November.

Young, C.E. (1984). Perceived water quality and the value of seasonal homes. *Water Resources Bull.,* 20(2): 163–166.

7

Opportunity Costs of Residential Best Management Practices for Stormwater Runoff Control

Hale W. Thurston

CONTENTS

Introduction

The effects of stormwater runoff on stream ecosystems are exacerbated by urbanization and the coincident increase in impervious surface in a watershed. Proliferation of impervious surfaces allows more water from rain events to reach a stream faster causing higher peak flows that can lead to stream alteration and habitat degradation. Where impervious surfaces prevent rainfall from infiltrating the soil, less water is available for groundwater recharge, which reduces stream base flow. Depending upon the land use in the watershed, nutrients and toxins can be scrubbed off roadways and parking lots and transported overland and through storm drains into waterways causing toxic loading of the stream.

Phase Two of the U.S. Environmental Protection Agency's (USEPA) National Pollution Discharge Elimination System (NPDES) stormwater regulations requires communities smaller than 100,000 residents to meet new criteria for stormwater runoff reduction. In many cases these smaller communities have no established stormwater utility, and are investigating alternatives for complying with these new, sometimes expensive requirements.

In a previous study, Thurston et al. (2003) hypothesized that a dispersed set of simple water retention technologies would control the negative impacts of runoff for an urban area at a lower cost than large engineering infrastructural projects. In that paper we suggest that a market-based tradable allowance mechanism for trading runoff reductions would be a practical and cost-effective method to assign dispersed runoff control throughout urbanized areas. However, we added several caveats to our analysis, not the least of which was ignoring opportunity costs of dedicating land area to the best management practices (BMPs). In this study we turn our attention to the estimation of those opportunity costs for the most prevalent land use and BMP types in our pilot study watershed. Specifically, we use hedonic modeling techniques to estimate residential homeowners' valuations of the land that makes up their parcel, and assume that the negative of this value represents a lower bound for the opportunity cost of a BMP. We then employ that value in a numerical exercise that compares two different incentive-based policies for stormwater control: tradable allowances and a fee-with-rebate approach.

This chapter proceeds as follows. The second section introduces the market mechanisms of interest, and discusses their appropriateness. In the third section we discuss the importance of the inclusion of opportunity costs in these market-based incentive policies, and how we estimate the opportunity costs in this study using the hedonic price estimation technique. The fourth section outlines how we apply the estimates from the section preceding it in a numerical example in the Shepherd Creek area of Cincinnati, Ohio. The last section concludes.

Economic Incentive Mechanisms

Fee and Rebate

Known as a Pigouvian tax, the optimal tax on pollution should be a direct tax equaling the marginal external damages caused by the pollution (Pigou, 1962). But directly taxing pollution is sometimes hard, especially when monitoring is difficult (such as with a nonpoint source) or when there are institutional barriers to imposing differing taxes on people in the same area. Fullerton and Wolverton (2003, 1999) note that when directly taxing the pollution is not an option, the policy maker can exploit a relationship the optimal pollution tax has with income tax and rebates on products that are related to, and relatively cleaner than, the polluting good. This relationship is derived mathematically by the authors, and describes the conditions whereby a uniformly applied increase in the local tax on income in conjunction with a rebate for some desirable behavior, act together to exactly mimic the Pigouvian tax. We do not pretend to achieve optimality in this chapter; rather we investigate the realism of a few ad hoc fee/rebate levels.

This type of policy is already in place in many municipalities in the United States (Doll, Scodari, and Lindsey, 1998; Doll and Lindsey, 1999). Unfortunately these programs are almost exclusively for commercial, not residential, properties and, where applicable, residential fees are too small to warrant a rebate. For example, monthly residential stormwater fees in Columbus, Ohio, St. Louis, Missouri, and Indianapolis, Indiana, are about $2.70, $0.24, and $1.25, respectively. Many agree that the existing programs have not encouraged the desired behavior because the fees and rebates are simply too low (Doll and Lindsey, 1999; Parikh et al., 2005). We show in the next section that if the fee and the rebate are high enough to make households reflect their true underlying preferences, based on our knowledge of the costs facing the residential property owner, a stormwater runoff reduction goal can be met using dispersed BMPs and a two-part instrument at a relatively low cost to the utility and the stakeholders.

Tradable Allowances

The cost-effectiveness of the tradable allowance approach to pollutant reduction in air sheds is well established in the literature (Tietenberg, 2000; Baumol and Oates, 1988; Eheart, 1980), and the SO_2 trading program in the United States has been operating successfully for several years. Watersheds differ from air sheds, however, in key aspects, such as the confinement to a channel, nonuniform mixing, and downstream accumulation, and these present new challenges for the establishment of tradable allowance systems. Watershed trading is not a new concept, but the specific application explored in this chapter is. For those interested in other applications, the USEPA's *Draft Framework for Watershed-Based Trading* (1996) provides an overview of some 20 tradable allowance programs across the United States. Several of these grew out of cooperative agreements with the USEPA, the Water Environment Research Foundation, and various local stakeholder groups. These programs focus on reducing concentrations of nutrients or toxics, and most rely upon an organizational effort similar to the USEPA's total maximum daily load (TMDL) process to drive stakeholder involvement.

Two necessary conditions for tradable allowance regimes to be cost reducing are that (i) transaction costs of such programs be no greater than the gains achieved and (ii) there be sufficient difference in abatement cost across parcel owners so that potential cost savings can be realized through market exchange of runoff control. With these conditions satisfied, a tradable allowance system can efficiently assign runoff control to dispersed locations and may avoid the larger cost of centralized approaches. I thank one anonymous reviewer for pointing out that in some cases the low-transaction cost condition may not be achieved. This will be true in some areas of the country, however, in those municipalities where an established stormwater utility has access to up-to-date geographic information systems (most do), citizens have access to the Internet, and a "clearinghouse" type of program such as that proposed by Woodward and Kaiser (2002) can be established, transaction costs will probably be low enough to take advantage of trading. Empirical testing of this theory is being investigated by researchers in the USEPA's Sustainable Environment Branch.

The usefulness of inclusion of opportunity costs can hardly be overstated in this application. We use the results of our opportunity cost estimation to inform a tradable allowance system much like those currently used in water quality trading programs around the country (USEPA, 1996), and much like the one specific to stormwater abatement that we describe elsewhere (Thurston et al., 2003).

Opportunity Costs

Under either of our proposed policy choices the costs to the homeowner must be known for the planner to be able to predict the success of the program with any accuracy. There are multiple components of cost facing each parcel owner: the construction of BMPs, their maintenance, and the opportunity cost of land taken out of other uses. A corner of the backyard has other utility-generating uses: gardens, sandbox, swing set, and so on, and so the homeowner dedicating an increasing proportion of the backyard to water detention incurs an opportunity cost. Together these form the marginal (or incremental) cost of abating a given unit of runoff.

If $CT = f(Q)$ is the total cost of abating runoff then the marginal abatement costs (MC) are given as:

$$\partial CT / \partial Q = MC$$

where Q is the volume of runoff abated. MC are assumed to increase monotonically with quantity abated such that

$$\partial MC\big/_{\partial Q} \geq 0$$

primarily due to increasing opportunity cost of land. That is, although the specific abatement technologies may exhibit decreasing costs to scale over the relevant size of the BMP, we assume that an individual property owner faces increasing opportunity costs for land taken out of other uses and dedicated to BMPs. A negative sign on the quadratic term for land area in our hedonic price curve estimation confirms this is the case with residential properties.

Making the standard assumption that property owners facing explicit costs for managing runoff from their properties will be cost-minimizers, we can predict the effects of changes in fee and rebate rates or allowance prices and in land use (e.g., increased impervious surface) on runoff flow and stream loading in the watershed.

Estimating Opportunity Costs of BMPs

It has recently been recognized that when considering other than engineering solutions to stormwater management, the inclusion of other than engineering costs is imperative, and that these costs are usually not known. For example, Sample et al. (2003) note that in the vast literature on costs of facilities for stormwater detention and retention, including costs of length of pipes, manholes, and the like, "data on BMPs are probably the least reliable." In an attempt to include opportunity costs in their life-cycle cost analysis of BMPs, the authors focus on a case study found in the work on sewer design done by Tchobanoglous (1981). Sample et al. (2003) use the discounted stream of benefits, based on real estate appraisals of land value forgone by dedicating land to BMPs. This approach is unique, and considers the time dimension of the opportunity cost, however, we choose a different approach based not on land value data from realtors, but extracted through hedonic analysis from market transactions. It is argued in the economics literature that hedonic estimation is superior to assessors' estimates because the hedonic method relies on actual (revealed preference) sales data, allows for marginal analysis, and is immune to the biases of an assessor (Leggett and Bockstael, 2000; Mooney and Eisgruber, 2001). In addition, inasmuch as multivariate regression separates out the constituents that make up the total price of the good in question we can focus on marginal effects. Future research might focus on the use of the Sample et al. (2003) method in conjunction with the hedonic method to more fully inform a policy about on-site residential BMPs and associated opportunity costs.

Hedonic models exploit the existence of heterogeneity in attributes that make up a given item that is traded in the market to estimate the portion each attribute adds to the value of the good. We estimate a hedonic price function that includes various elements of properties sold in Hamilton County, OH, in postal zip codes near the Shepherd Creek area. Estimation of hedonic price function for housing is commonplace (Haab and McConnell, 2002).

Econometrically, estimation is fairly straightforward. Covariates usually include things such as number of bathrooms, fireplaces, the existence of central air conditioning, and so on. Some neighborhood characteristics are usually also included such as school quality and proximity of parks. Our focus is on the value homeowners place on their yards; if square footage is dedicated to an on-site stormwater detention facility such as a sand trench or rain garden it is lost to other uses. To get at this we include a covariate "NETLOT," which is the area of a property alone (lot size minus footprint of house) in square feet.

We estimate the hedonic price function

$$p = h(z, \alpha)$$

where p is the price of the house, z is a vector of attributes such as those mentioned above, and alpha is a vector of parameters that define the shape of the price function. We assume the standard approach that the household budget constraint is given by

$$b = h(z) + x$$

where x is all other goods, and maximizing utility subject to the budget constraint implies optimal conditions for each attribute

$$\frac{\partial u(x, z : \beta)}{\partial z_l} = \lambda \frac{\partial h(z)}{\partial z_l}, l = 1, \dots L$$

If equilibrium conditions hold in the housing market the value owners place on the attribute "NETLOT" is expressed as

$$\hat{wtp} = \Delta z_l \frac{\partial h(z)}{\partial z_l}$$

where z_i is the covariate "NETLOT" and the result is the amount of land dedicated to a BMP multiplied by, in the case of linear estimation of the hedonic price function, the estimated coefficient of the land variable. The log-linear estimated coefficients are interpreted as percentage of parcel sale price due to the attribute in question. It follows that, if land is dedicated to a BMP, the negative of the above represents the opportunity cost of land in the BMP.

Data

Publicly available data on over 100,000 house sales from the period 1998–2001 were downloaded from the Hamilton County, Ohio, County Auditor's website. Data included 103 columns of information, including sale price, parcel numbers, address, name of buyer, date of the sale, and school district among others. Also included is information on attributes of the property such as number of bathrooms, square footage, central air conditioning, and the like. Only residential, nonvacant properties that were sold in zip codes adjacent to the pilot project site were used, and we culled incomplete and obviously erroneous entries resulting in 24,218 observations. Variables used in ordinary least square (OLS) estimation are listed in Table 7.1, along with their means and definitions.

Results

Estimation results are presented in Table 7.2. A quadratic term for square footage of lawn is included to reflect decreasing marginal utility. We test the choice of the linear versus log-linear specification using the Box and Cox (1964) test statistic under the null hypotheses that the two are equivalent, and we reject the null hypothesis and conclude the models are not observably equivalent, and determine the log-linear model provides better estimates. We use the Box–Cox test because the transformation of the dependent variable renders R-squared values incomparable (Griffiths, Hill, and Judge, 1993). The coefficients estimated in the log-linear specification are interpreted as percentage of parcel sale price due to the attribute in question. Regression results in each yield coefficient signs as expected and all estimated coefficients are significant at the 95% confidence interval.

The coefficient of most interest for our purposes is that on NETLOT. The regression results tell us that the part worth an additional square foot of yard is worth 1.17×10^{-6} times the sale price of the house. Our regression is run on data on houses that have sold in zip codes in and adjacent to our pilot study watershed. Because we do not have enough observations on sales in the pilot area to return statistical significance, we take the average of the regression sample to represent the average house in the pilot study

TABLE 7.1

Variables Used in Regression Analysis

Variable n = 24218	Mean	Definition
HALFBAT	0.44322	Number of half bathrooms
FULLBAT	1.5114	Number of full bathrooms
SFTFIN	1551.8	Area of the house finished in square feet
NETLOT	0.32064	Area of property alone (lot size minus footprint of house) in square feet
NETLOTS	0.75029	Net square footage of property squared
SCHRANK	15.159	School rank in Hamilton County based on statewide aptitude test pass percent[a]
FIRE	0.46061	Number of fireplaces
COND	3.4385	Condition rating assigned by realtor, 1–5
SALEPRIC	107,370	Sale price of the property in current dollars
Cool	0.72108	Dummy variable that takes the value 1 if central air conditioning, 0 otherwise

[a] Skertic, M. *Cincinnati Enquirer* (November 30, 1997) http://enquirer.com/schools/highlights.html

TABLE 7.2

Regression Results

	Parameter Estimates Standard Errors in Parentheses	
Characteristic	Linear	LOG-LIN
HALFBAT	11168	0.11511
	(642)	(0.00597)
FULLBAT	14065	0.11219
	(693)	(0.00645)
SFTFIN	44.7	0.000256
	(0.745)	(0.00000693)
NETLOT	0.278	1.17×10^{-6}
	(0.0137)	(1.28×10^{-7})
NETLOTS	-9.39×10^{-8}	-4.44×10^{-13}
	(7.88×10^{-9})	(7.33×10^{-14})
SCHRANK	−1908	−0.0169
	(56.1)	(0.000522)
FIRE	15800	0.155
	(655)	(0.00609)
COND	6820	0.115
	(416)	(0.00387)
COOL	8280	0.199
	(723)	(0.00672)
CONSTANT	294	10.4
	(2200)	(0.0204)
R^2	0.5464	0.4567

area. The mean value from Table 7.1 is $107,370 making the estimated part worth 12.6 cents. The interpretation of the linear coefficient is of course much simpler as the coefficient is the part worth, in this case 28 cents. How do we translate this into opportunity costs? Either of the BMP technologies available to the landowner in the pilot area will require a certain amount of yard space to be taken out of other uses. To get an idea of the scale of the proposed BMPs, it is not unrealistic to assume that, in the case of a small-scale residential BMP or porous pavement, the footprint (i.e., the square footage of the BMP on the property) is the same as the volume capacity. That is, a 1.0 m (3.3 ft) deep (3-ft deep sand trench BMP would have to measure 6.1 m by 3.0 m (20 ft by 10 ft) to capture 5.7 m³ (200 ft³) of stormwater runoff (to accommodate the sand filler with typical porosity of around 35%, this is thrice the necessary empty void space).

Application to a Case Study

Cincinnati's Mill Creek and its tributaries occupy some 44,000 ha (170 square miles) in 23 municipalities in the counties of Hamilton and Butler, Ohio. Mill Creek is considered by the Ohio Environmental Protection Agency to be the most polluted waterway in the state (OEPA, 1994). Stormwater runoff is a major direct and indirect contributor to the pollution of Mill Creek. To investigate the usefulness of our cost estimates for an incentive program we proceed as follows: we delineate the boundaries of the Shepherd Creek subwatershed using established hydrology. The subwatershed is divided into parcels, then characteristics such as soil type, land use, and parcel boundaries are entered into an ArcView GIS project. Figure 7.1 shows the area in detail, illustrating the headwaters, parcels, different soils, and impervious surfaces.

We chose this section of Mill Creek for our study area because of the diverse land use, topography, and soil types. Mount Airy Forest is a protected area that occupies a large piece of the study area, and proposed and ongoing subdivision developments provide us with heterogeneous land use and the opportunity to research a dynamic system. The impact of a rainfall event on the study area was determined using the Natural Resource Conservation Service's Technical Release (TR-55) bulletin, which provides the methodology to calculate stormwater runoff volume for a given storm event. The storm event we use in our model is the 1.5-year storm. In the case of Hamilton County, Ohio, that rainfall amount is 3.12 cm (1.23 in.). Our choice of the illustrative rainfall event is based on expert opinion on low-impact development design such as outlined in Prince George's County, Maryland (2000). Use of these storm events as

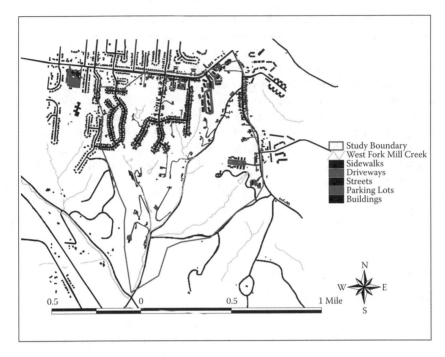

FIGURE 7.1
Impervious area in Shepherd Creek.

opposed to using larger, more infrequent, storms goes back to Wolman and Miller (1960) who conclude that smaller frequent floods have greater impact in the long run on sediment transport and therefore on habitat alteration than do large infrequent floods.

It is assumed the stormwater utility has either determined the ecological limits of streams and sewers in the relevant watersheds, or, as is the case in our exposition, imposes a reasonable ad hoc reference level (such as runoff in predevelopment conditions), and grandfathers each parcel owner an allowed flow of water from each parcel. Runoff volume in excess of the grandfathered amounts will be subject to the stormwater fee.

Rain event runoff is calculated using the TR-55 methodology for each parcel in the watershed, taking into account the existing impervious surface. We identify this as the postdevelopment runoff. Runoff is calculated again using the same rain event, but modeled in the absence of the impervious surface and with a forested land cover. This result is identified as the predevelopment runoff. Excess stormwater runoff then, is the postdevelopment runoff minus the predevelopment runoff. We use the event-based simulation to keep illustration of our proposed policy simple. Ongoing research includes the application of the U.S. Department of Agriculture's

AnnAGNPS continuous runoff model. The main improvement this offers is the explicit acknowledgment that soils become saturated during periods of frequent storms.

Mathematical calculation of the property owners' runoff responsibilities is implemented in Microsoft Excel, and then exported to ArcView for visualization. Real-time linking via structure query language (SQL) connection or other means is possible and may be used in the future as a tool by a stormwater utility, for example. An ArcView representation of excess stormwater runoff caused by the design storm in Shepherd Creek is shown in Figure 7.2.

We created a spreadsheet database that contains physical and hydrological features of parcels in the Shepherd Creek subwatershed. The database also includes the cost of different BMPs as a function of their capacity. The costs include construction and the costs estimated as opportunity costs from our hedonic price function. Each property has associated with it a parcel identification number, a land-use type classification (multifamily, industrial, etc.) a soil type, and quantity of excess stormwater runoff in cubic feet as calculated from the TR-55 methodology outlined above.

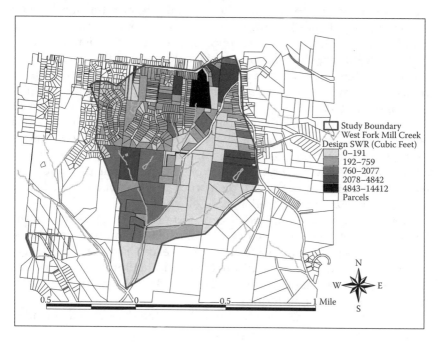

FIGURE 7.2
Excess stormwater runoff.

Numerical Example

We use 402 single-family residential parcels from the Shepherd Creek subwatershed delineated above, in our numerical example. The soil is of hydrological soil group (HSG) B, C, or D and the terrain ranges from forested to grassy to impervious. Parcel owners in the case study area were assumed to choose the most suitable BMP for their land use and soil type from among a variety of BMPs that exist for different applications.

For the computational exercise, the type of BMP, concomitant detention cost (DC) functions, and inverse cost functions were assigned to parcels in the subbasin based on land use and soil type as follows:

- Parcels with residential land use and HSG B are assumed to employ sand filters. The cost function is: $DC_{ResB} = 26.6Q^{0.64} + 0.126Q$.

- Residential parcels with HSG C or D are assumed to use rain gardens/grassed swales, the relevant cost function of which is $DC_{ResC} = 4.94Q + 0.126Q$,

where Q is the quantity of stormwater runoff detained in cubic feet, and the second term in each is the log-linear estimated opportunity cost of land. We choose not to include the quadratic term estimated because although it is statistically significant, it is not economically significant at the BMP sizes with which we are concerned. The cost functions are modified from Schueler (1987) and Heaney, Sample, and Wright (2002).

For this exercise, some properties, such as public schools and public rights of way, were not included. Although these do contribute a significant amount of stormwater runoff and toxic loadings to streams, we do not include public agency responsibility in the present analysis. The area of the roads in this watershed is 4% of the total land area. According to TR-55, the contribution to the runoff volume from rain that falls directly on the roads is 15% of the total runoff volume. This is partially mitigated by extant requirements for BMP construction for public infrastructure (CAGIS, personal communication, June 20, 2001).

Average excess runoff for the design storm from the properties in the Shepherd Creek study area is 8.27 m³ (292 ft³) of stormwater. Some small or undeveloped properties have virtually no excess runoff whereas large multifamily apartment-covered properties with multiple driveways and rooftops can have as much as 85 to 142 m³ (3,000 to 5,000 ft³) of stormwater runoff due to the design storm.

First we calculate what the cost would be, in the absence of any market incentives, to control the subwatershed's excess stormwater runoff. To calculate the cost of a dispersed set of BMPs to store on site all of the excess runoff from a storm event of 3.12 cm (1.23 in.), the design storm, we assign the appropriate least-cost BMP technology on a parcel-by-parcel basis in our small case area and solve each landowner's cost equation. This calculation presumes that the parcel owner is responsible for all the runoff over and above that which would result if the parcel were in its undeveloped state. This in effect is a command-and-control regime with no rebate and the water detained with this constraint is 2,344 m³ (82,767 ft³), at an average cost of $950 per homeowner. Considering only construction costs of BMPs results in an average cost of $4.62 per cubic foot of stormwater runoff detained via BMPs over all properties in the study area. We now calculate including the opportunity cost of land, estimated from the hedonic price functions noted in the cost functions above, the log-linear case is $0.126 per ft². Assuming again that command-and-control policy is implemented that causes all parcel owners to use BMPs to manage all excess stormwater runoff, the cost is $6.49 per ft³, at an average cost per homeowner of $1,337. We would assume costs to be higher for all price levels of allowances due simply to the increased cost of BMPs (recognition of opportunity costs).

Incentives I: Fee and Rebate

Now we apply a two-part tariff that makes the construction of BMPs voluntary, but payment of the stormwater fee mandatory. Armed with the cost functions for the homeowners in the parcel area (and with a visualization of that in the GIS platform) the stormwater utility can use any spreadsheet program to try various combinations of fees and rebates to determine what combination thereof will achieve the predetermined hydroecological limit on excess runoff and at what price. In our Excel spreadsheet each parcel is defined by various columns. Each has associated with it the quantity of excess stormwater runoff responsibility (as calculated above), soil type, land use, the appropriate BMP technology, the BMP cost function, and the fee and rebate facing the parcel. At different user-entered fee and rebate levels and based on simple "if/then" analysis of the costs, the parcel is assigned either a BMP and rebate or the cost of the fee. Summation across the subwatershed's 402 parcels yields the results presented in Table 7.3.

For Shepherd Creek we give examples in Table 7.3 showing the per-unit cost of detention assuming some different levels of fee and rebate, with and without the inclusion of opportunity cost of land for residential BMPs. The very small, or even negative (designated in parentheses), cost to homeowner figures come about because some homeowners can build a BMP sufficient to control all their runoff for significantly less than the amount of the fixed rebate. Net Utility Revenue ignores almost all the costs a real utility would incur; here it is simply the difference of revenues from fees minus rebates given out. The total excess runoff in the subwatershed is 2,344 m³ (82,767 ft³). The stormwater utility, it is assumed, would consult

TABLE 7.3

Per-Unit Cost of Detention Using Fee and Rebate

Scenario	No Opportunity Costs			With Opportunity Costs		
	Cost per Cubic Foot Detained (to Average Homeowner) ($/ft³)	Quantity Detained (ft³)	Net Utility Revenue ($)	Cost per Cubic Foot Detained (to Average Homeowner) ($/ft³)	Quantity Detained (ft³)	Net Utility Revenue ($)
No rebate	4.62	82,767	N/A	6.49	82,767	N/A
			Command and Control			
Scenario			**Fee = $1000**			
$1000 rebate	7.97	38,809	136,000	11.45	27,967	191,000
$1500 rebate	2.41	61,283	(130,500)	4.04	47,311	(48,000)
$2000 rebate	(0.55)	71,864	(368,000)	0.48	58,689	(286,000)
Scenario			**Fee = $500**			
$1000 rebate	2.79	38,809	(65,000)	4.26	27,967	(10,000)
$1500 rebate	(0.87)	61,283	(331,500)	(0.21)	47,311	(249,000)
$2000 rebate	(3.35)	71,864	(569,000)	(2.94)	58,689	(487,000)
Scenario			**Fee = $1500**			
$1000 rebate	13.15	38,809	337000	18.64	27,967	392,000
$1500 rebate	5.69	61,283	70,500	8.29	47,311	153,000
$2000 rebate	2.24	71,864	(167,000)	3.91	58,689	(85,000)

with hydrologists and ecologists to set its limit on excess stormwater run-off. Obviously the quantity controlled will be positively correlated to the size of the rebate. But inasmuch as the quantity controlled is independent of the fee charged, the utility will be hard-pressed to operate in the black long with the superhigh rebates required if the fees are small. These fees and rebates are colossal compared to those currently offered in stormwater programs around the country, although they would probably only be assessed once during the lifetime of a BMP whereas the fees currently being charged by utilities are in perpetuity. In this subwatershed the most stormwater runoff the utility can cause to be detained without net revenues being negative is 1,340 m^3 (47,311 ft^3) if one accounts for the opportunity costs. If the utility misspecified cost functions and did not include opportunity costs it might estimate it could control 1,735 m^3 (61,283 ft^3) of stormwater runoff and remain in the black.

Incentives II: Tradable Allowances

Now we look at the imposition of the policy where the homeowner chooses to buy a credit or allowance from the stormwater utility or build an appropriately monitored and sized BMP of the correct technology on their parcel. The trading proposed here, as is the case in most water trading programs around the nation, is a "clearinghouse" approach. For a good overview of this and other market structures for water trading programs in the United States see Woodward and Kaiser (2002). In part to reduce transaction costs, instead of freely traded allowances on an open market, an authority, such as a stormwater utility, sets the price of allowances and this price is known to all participating parties. Parcel owners then choose to pay the fee or build a BMP to control their runoff. Those in the subwatershed who build the BMPs will be partially or wholly reimbursed by the stormwater utility from proceeds of allowances sold to parcel owners who choose not to build BMPs.

The sample is the same 402 residences in the Shepherd Creek area. The appropriate BMPs, and therefore appropriate cost functions (with the opportunity costs), are the same as in the fee and rebate case. The only difference here in the "fee" is the purchase of an allowance and is voluntary, but is functionally tied to the excess runoff of the property. The alternative is to pay no fee, but install a sufficiently sized BMP on the property. We are concerned mostly with retrofit opportunities for established communities here, so we do not consider the alternative that is currently incorporated in newer housing developments: a central stormwater control facility. Elsewhere (Thurston et al., 2003), we found that the per-unit cost of dispersed on-site retention compared favorably to that of a large-scale detention facility. The model of the parcel owner's choice is again carried out on an Excel spreadsheet. Each parcel is faced with a user-entered allowance price and the owner will choose to buy the allowance when the price is below the cost of building a BMP according to the cost function facing her. When the cost of the allowance is above the cost of the BMP the homeowner builds the BMP. When making this decision, the homeowner will, of course, want to take into consideration the opportunity cost of the BMP. A more thorough overview of this unique stormwater control application, but without the opportunity costs, is presented in Thurston et al. (2003). Applying the decision matrix presented above yields the varied results presented in Table 7.4.

These calculations were made assuming that the parcel owner chooses to retain *all* the excess runoff, or to retain *none* of it and purchase allowances. We can obtain values higher than the allowance price because the figure has quantity detained as the denominator. When we do not consider the opportunity costs of the BMPs, even $8.00 per ft³ price of allowance is considered high by the property owners, reflected in the miniscule number who bought them (almost all the excess runoff is detained, signifying many BMPs were built). Only somewhere between $5.00 and $4.00 is there a switching point when the allowances become almost uniquely attractive (by $2.50 no BMPs are built, because all homeowners buy allowances). But when we consider the opportunity costs to the parcel owner of the BMPs, allowances naturally become more attractive.

TABLE 7.4

Per-Unit Cost of Detention Using Tradable Allowances

Scenario Allowance price/cubic foot (not) detained	Average Cost per Cubic Foot Detained ($/ft³)	Quantity Detained (ft³)	Allowance Revenue ($)	Average Cost per Cubic Foot Detained ($/ft³)	Quantity Detained (ft³)	Allowance Revenue ($)
	Without Opportunity Costs			With Opportunity Costs		
$8.00 Allowances	4.62	82,574	1,544	7.16	70,060	101,656
$5.00 Allowances	4.65	81,721	5,230	12.98	29,784	264,915
$4.00 Allowances	24.01	13,517	277,000	27.86	11,706	284,244
$2.50 Allowances	N/A	0	206,917	N/A	0	206,918
Command-and-Control	4.62	82,767	0	6.49	82,767	0

Conclusions

This study is part of an ongoing effort to fully describe a realistic market-based mechanism to alleviate the problems caused by increased stormwater runoff in urban and urbanizing areas. As impervious surfaces grow relative to natural landscapes the problem will increase, and municipal authorities will look for ways to deal with it that are practical both from a political standpoint and ecologically effective. We have outlined how employment of BMPs in a subwatershed would reduce stormwater runoff substantially, but the need to prove our assumptions about the policy that would bring this about are not unrealistic. To this end we show that inclusion of the opportunity cost of residential land dedicated to particular stormwater control practices increases the overall cost of stormwater control in a pilot study area. Using hedonic modeling techniques we have estimated the value parcel holders place on their land area, and used that information in our incentive schemes.

In some cases we show that the incentive schemes are more cost-effective than command-and-control, however, the evidence is weaker when the cost functions more realistically include the opportunity cost. Nevertheless, command-and-control costs are presented primarily as a hypothetical benchmark inasmuch as it is probably unrealistic to think that a stormwater utility would have the regulatory might to impose such a regulation. Further research includes the collection of experimental auction data from homeowners to validate some of our assumptions about decision making we model, and the actual implementation of BMPs for the purpose of measuring effectiveness. With this implementation will come the opportunity to measure heretofore ignored, as noted by one anonymous reviewer, costs of operation and maintenance. These costs, although we assume them to be small due to the low-tech nature of the BMPs, should be considered in an overall assessment of the cost-effectiveness of a dispersed detention program.

Acknowledgment

This chapter is reprinted from Thurston, Hale. W. (2006). Opportunity costs of residential best management practices for stormwater runoff control. *Journal of Water Resources Planning and Management*, 132: (2): 89–96. American Society of Civil Engineers. With permission from ASCE.

References

Baumol, W.J., and Oates, W.E. (1988). *The Theory of Environmental Policy*. New York: Cambridge University Press,

Box, G. E. P., and D.R. Cox. (1964) An analysis of transformations, *J. R. Statist. Soc.* Series B, 26(2): 211–243

Doll, A., and G. Lindsey. (1999). Credits bring economic incentives for onsite stormwater management. *Watershed Wet Weather Tech. Bull.*, 4(1): 12–15.

Doll, A., P.F. Scodari, and G. Lindsey. (1998). Credits as economic incentives for on-site stormwater management: Issues and examples. *Proc., U.S. Environmental Protection Agency National Conference on Retrofit Opportunities for Water Resource Protection in Urban Environments*, USEPA, Chicago, 113–117.

Eheart, J.W. (1980). Cost-efficiency of transferable discharge permits for the control of BOD discharges. *Water Resources Res.*, 16(6): 980–986.

Fullerton, D., and A. Wolverton. (1999). The case for a two-part instrument: Presumptive tax and environmental subsidy. In A. Panagariya, P. Portney, and R. Schwab (Eds.), *Environmental and Public Economics: Essays in Honor of Wallace E. Oates*. Cheltenham, UK: Edward Elgar.

Fullerton, D., and A. Wolverton. (2003). The Two-Part Instrument in a Second-Best World, Working paper Series, NCEE, USEPA.

Griffiths, W.E., R.C. Hill, and G.G. Judge. (1993). *Learning and Practicing Econometrics*. New York: John Wiley and Sons.

Haab, T.C., and K.E. McConnell. (2002). *Valuing Environmental and Natural Resources: The Econometrics of Non-Market Valuation*. Northampton, MA: Edward Elgar.

Heaney, J.P., D. Sample, and L. Wright. (2002). Costs of Urban Stormwater Control. EPA Report No. EPA-600/R-02/021; NTIS No. PB2003103299. National Risk Management Research Laboratory, Cincinnati, OH: U.S. Environmental Protection Agency.

Leggett, C.G., and N.E. Bockstael. (2000). Evidence of the effects of water quality on residential land prices. *J. Environ. Econ. Manage.*, 39: 121–144.

Mooney, S.N., and L.M. Eisgruber. (2001). The influence of riparian protection measures on residential property values: The case of the Oregon Plan for salmon watersheds. *J. Real Estate Fin. Econ.*, 22(2/3): 273–286.

Ohio Environmental Protection Agency (OEPA). (1994). Biological and Water Quality Study of Mill Creek and Tributaries, Ohio Environmental Protection Agency Technical Report SWS/1993-12-9, Columbus.

Parikh, P., M.A. Taylor, T. Hoagland, H. Thurston, and W. Shuster. (2005). Application of market mechanisms and incentives to reduce stormwater runoff: An integrated hydrologic, economic and legal approach. *Environ. Sci. Policy*, 8: 133–144.

Pigou, A.C. (1962). *The Economics of Welfare*, 4th ed. London: Macmillan.

Prince George's County, Maryland, Department of Environmental Resources. (2000). *Low-Impact Development: An Integrated Design Approach*, EPA 841-B-00-003, January.

Sample, D.J., J.P. Heaney, L.T. Wright, C. Fan, F. Lai, and R. Field. (2003). Costs of best management practices and associated land for urban stormwater control. *J. Water Resources Plan. Manage.*, 129(1): 59–68.

Schueler, T.R. (1987). *Controlling Urban Runoff: A Practical Manual for Planning Designing Urban BMPs*, Washington, DC: Washington Metropolitan Water Resources Planning Board.

Skertic, M. (1997). Home income predictor of school test results. *The Cincinnati Enquirer*, 30 November.

Tchobanoglous, G. (1981). *Wastewater Engineering: Collection and Pumping of Wastewater*. New York: McGraw-Hill.

Thurston, H.W., H.C. Goddard, D. Szlag, and B. Lemberg. (2003). Controlling stormwater runoff with tradable allowances for impervious surfaces. *J. Water Resources Plan. Manage.*, 129(5): 409–418.

Tietenberg, T. (2000). *Environmental and Natural Resource Economics*. Reading, MA: Addison-Wesley.

USEPA. (1996). *Draft Framework for Watershed-Based Trading*, Office of Water, EPA 800-R-96-001, May.

Wolman, M.G., and Miller, J.P. (1960). Magnitude and frequency of forces in geomorphic processes. *J. Geol.*, 68(1): 54–74.

Woodward, R.T., and R.A. Kaiser. (2002). Market structures for U.S. water quality trading. *Rev. Agric. Econ.*, 24: 366–383.

8

At the Intersection of Hydrology, Economics, and Law: Application of Market Mechanisms and Incentives to Reduce Stormwater Runoff

Punam Parikh, Michael A. Taylor, Theresa Hoagland,
Hale W. Thurston, and William Shuster

CONTENTS

Introduction

Excess stormwater runoff, caused by urbanization and the increased proportion of land area under impervious surfaces, has negative impacts on both terrestrial and aquatic ecosystems. The proposed interdisciplinary approach involves providing incentives for the construction of small-scale best management practices (BMP) throughout a small watershed, leading to a cost-effective means to control stormwater runoff, and partially restoring a more natural hydrological regime to a watershed area.

Communities nationwide are becoming more aware of the damage to the environment and personal property that stormwater runoff causes. This is due in part to the recognition that urban sprawl and increased impervious surfaces cause increased runoff, and in part because many previously unregulated communities are subject to the U.S. Environmental Protection Agency's recently promulgated National Pollutant Discharge Elimination System (NPDES) stormwater measures.[*] Controlling excess stormwater runoff poses some challenging issues: stormwater runoff is not defined as a pollutant, nor is the source of runoff well defined. The stated objectives of excess stormwater management typically are to provide for drainage from upland communities, to minimize downstream impacts of upstream development, and to balance the environmental and social impacts of the drainage infrastructure. Yet, these objectives clash more often than not.

It has been suggested that market mechanisms similar to pollutant trading could be used to encourage the use of distributed (i.e., as opposed to centralized) stormwater control measures. The proposed management approach involves providing incentives for the construction of small-scale best management practices (e.g., infiltration trenches, small detention ponds) throughout a small watershed, leading to a cost-effective means to control stormwater runoff, and partially restoring a more natural hydrological regime to a watershed area. Recognizing that a uniform generalized solution is seldom effective (Booth and Jackson, 1997), we look at a range of possible management scenarios from the hydrological, economic, and legal standpoints. The multifaceted nature of the problem calls for the kind of interdisciplinary approach we take in this research. This chapter first provides some background on the hydrology of stormwater runoff, the economic aspects of proposed solutions for dealing with the runoff, and the legal responsibilities and recourse communities have in dealing

---

[*] National Pollutant Discharge Elimination System—Regulations for Revision of the Water Pollution Control Program Addressing Storm Water Discharges; Final Rule, 64 Fed. Reg. 68,722 (Dec. 8, 1999).

with stormwater runoff. The next section presents four implementation scenarios and discusses their relative attractiveness from each discipline's perspective. The last section synthesizes these perspectives and concludes the chapter.

Background: Stormwater Runoff

Hydrological Aspects

There are two basic runoff formation processes: pervious surfaces that infiltrate water eventually become saturated and the excess precipitation runs off; or a pervious surface is replaced with an impervious surface that does not allow the infiltration of water. Therefore, precipitation is more completely converted to runoff, which also initiates within a shorter time frame. In urbanized areas, the increase in stormwater runoff is a direct result of the proliferation of impervious surfaces and a concomitant decline in natural sinks for storm flows (Hey, 2001). All properties within a watershed contribute stormwater runoff to some extent. Each parcel contains different combinations and arrangements of types of impervious surfaces, which affects the amount of runoff produced and the time at which it is delivered to other parts of the catchment.

The intensity and duration of a precipitation event is the major determinant of runoff volume (Church, Granato, and Owens, 1999). The stormwater infrastructure has historically provided the capacity to handle stormwater from small storm events (i.e., 2–10 year events) and large events (i.e., 50–100 year events) and these systems are known as convenience and emergency systems, respectively (Walesh, 1989). The convenience system collects stormwater from downspouts, street inlets, and other portals, which is then conveyed to treatment plants; more generally, these flows are piped to streams and their tributaries. The emergency system attempts to address larger flows that result from the more infrequent, yet higher-volume stormwater runoff events, and relies on streets, roadways, and low-lying areas to act as open-channel conveyances and temporary detention basins, respectively. The terminal receptor for these flows is also typically a stream. These urbanized conditions route stormwater to streams in less time and in greater quantities, causing higher peak flows that lead to stream degradation and habitat alteration. Excess stormwater runoff has historically had a degrading impact on both terrestrial (Putnam, 1972; Johnson and Sayre, 1973; Cairns, 1995; Jauregui and Romales, 1996; Carlson and Arthur, 2000) and aquatic (Klein, 1979; Neller, 1988; Booth and Jackson, 1997; Swanson et al., 1998) ecosystems.

Economic Aspects

Traditional command-and-control regulations set uniform standards for all sources, with the most common being technology- and performance-based standards. Technology-based standards dictate the manner in which individuals must comply with the regulation. A performance standard sets a uniform control target for all sources that allows for some flexibility in the choice of abatement technology utilized. Rather than requiring individuals to meet uniform control targets, market-based instruments persuade individuals to expend equal marginal control costs. When the cost of controlling emissions or runoff differs across individuals, equalizing marginal control costs ensures that the overall target is attained at the lowest aggregate cost. A larger portion of the abatement burden is allocated to individuals with relatively lower abatement costs. The interested reader can refer to Appendix A for more detail and graphical illustration of this and subsequent control mechanisms.

Allocating the abatement burden among sources to achieve the lowest overall control costs can be accomplished through either a Pigouvian tax or through an allowance market. A Pigouvian tax is a price-based instrument that is a charge collected by a regulatory agent and that is levied on each unit of runoff from an individual's parcel. (See Chapter 7.) An allowance market is a quantity-based instrument that restricts allowable levels of runoff but permits the transfer of these allowances through free trade. The application is shown graphically in Appendix A, in this chapter.

Allowance markets can also achieve the lowest cost allocation of abatement among individuals, but without the same abatement cost information requirement. Under this quantity-based approach, each individual is required to have an allowance for the runoff that leaves his or her property. Each allowance defines the quantity of runoff that the individual is allowed to have exit the parcel. This is no different than saying that each allowance defines the amount of abatement that each individual is required to provide. The total number of allowances issued by the control agency must equal the total allowable level of runoff for the watershed. Finally, the individuals are allowed to buy and sell allowances to other individuals within the market. This permits the allowances to be reallocated to the individuals with the highest abatement costs, which results in the equalizing of marginal abatement costs and the low cost allocation of the abatement burden.

The use of market mechanisms for pollution control is attractive from a theoretical standpoint, and boasts measurable successes in air pollution trading markets. Less widespread, but gaining credence is the use of markets to create incentives for water quality improvements. Using markets to control excess stormwater runoff, a water *quantity* issue, is a novel approach that has, as we detail in this chapter, potential for successful application. Market-based instruments place an economic value on reducing excess stormwater runoff, and provide financial incentives for individuals to identify and adopt lower-cost control technologies. As with all potential applications, transfer from theory to application requires attention to details of "real-world" obstacles for effective implementation.

In practice, allowance markets have been gaining acceptance in a wide variety of applications. Perhaps the best known of these is the SO_2 market. Traded on the Chicago Board of Trade, the market in sulfur dioxide emissions has been a success in meeting overall emissions reduction targets at lower than expected attainment costs. Although a large degree of the markets' success can be attributed to its remaining faithful to theoretical design, this can also be criticized as its weakness. An underlying assumption of emissions trading theory is that the pollutant is uniformly mixing and that the point of discharge does not affect the degree of environmental impact. In reality spatial characteristics of discharge do have an effect on environmental impact. For example, Midwest emissions tend to concentrate in the Northeast causing environmental damage in that region (U.S. GAO, 2000; New York State Department of Environmental Conservation, 2004; Green Nature, 2002).

The use of trading markets for the control of water pollution is currently being encouraged, with 37 programs in the development or implementation stage in the United States (Environomics, 1999). Including nonpoint pollution within these markets necessitated several key modifications in the design of allowance markets due to its diffuse nature. Monitoring individual contributions can be prohibitively costly, loadings are in large part driven by random weather events, and uncertainty exists regarding the effectiveness of pollution abatement controls (Tomasi, Segerson, and Braden, 1994). Nonpoint sources of water pollution have not been directly regulated under the NPDES program of the Clean Water Act partly because of the difficulty of identifying individual contributions.[*] As a result, existing allowance markets have deviated from traditional design by (i) including nonpoint sources on a voluntary basis, and (ii) monitoring trades based on the adoption of abatement technology (e.g., best management practices) rather than observed performance. The latter modification has led to the expanded use of trading ratios, which we define and discuss in the next section. Within these markets the trading ratio deals not only

---

[*] The Clean Water Act's NPDES program focuses on permits for discharges of pollutants. 33 U.S.C. §1342(a) (2001). "Discharge of a pollutant" is defined in the Clean Water Act as "any addition of any pollutant to navigable waters from any point source." 33 U.S.C. §1362(12) (2001). "Point source" is defined as "any discernible, confined and discrete conveyance, including but not limited to any pipe, ditch, channel, tunnel, conduit, well, discrete fissure, container, rolling stock, concentrated animal feeding operation, or vessel or other floating craft, from which pollutants are or may be discharged." 33 U.S.C. §1362(14) (2001). Nonpoint sources, by nature, are not easily discernible nor do they flow from a confined or discrete conveyance and are therefore not subject to the NPDES program.

with differential spatial impacts of emissions, but also the uncertainty of the relationship between the estimated reductions from individual best management practices and actual emissions.

The use of market mechanisms, such as allowance markets, will require further modification to the traditional design of these instruments. For example, as with nonpoint sources the dispersal nature of stormwater runoff introduces questions of how to monitor and enforce trade within allowance markets. Stormwater runoff also adds the additional concern of not being a regulated pollutant, introducing new concerns regarding the ability to create appropriate legally enforceable incentives.

Legal Aspects

Although all properties within a watershed contribute to stormwater run-off, the property owner's rights and obligations associated with storm-water runoff vary depending on property type (commercial, residential, governmental), location (urban or rural), and controlling Federal or state law. Here, we focus on the Federal Clean Water Act's programs to reduce pollutants carried by stormwater, state water law as it relates to drainage and the right to dispose of excess water, and differences between trading to increase versus decrease water quantity.

Stormwater is addressed by the Federal Clean Water Act (CWA) under Section 402, which requires point source dischargers of pollutants to obtain an NPDES permit. Stormwater itself is not a pollutant as defined by the CWA, however, it often contains pollutants and so several sources of stormwater are considered point sources subject to the CWA. Those sources include certain industrial sectors, such as construction, publicly owned treatment works (POTWs), and municipal separate storm sewer systems (MS4s).

Stormwater presents a particular problem in older cities with combined sewer and sanitary systems (CSS). Combined systems are designed to collect and treat wastewater and stormwater together. When large storm events occur, many systems do not have the capacity to handle the excess stormwater and are forced to discharge this mix of stormwater and untreated wastewater directly into rivers or streams. When this occurs, the POTW is frequently forced to violate its NPDES permit because the permit does not allow wastewater to be discharged into waterways with-out treatment.

In addition to pollution law, relevant state water law deals directly with the drainage of diffused surface water (which includes stormwater). Water law is very complex and state-specific, but three general rules determine liability for damages caused by changes made to the natural flow of dif-fused surface waters. The majority of states follow the "common enemy" rule, which allows a landowner to protect his or her property from diffused water without regard to the consequences. The "natural servitude rule" (also called the "natural flow" or "civil law" rule) requires the lower landowner to accept the natural flow of water from the upper land, but prevents the upper landowner from changing the natural drainage flow such that it increases the burden on the lower land. There are exceptions to both, but some states use a third rule, which is "reasonable use." The reasonable use rule allows landowners to divert or change drainage water in any way that does not unreasonably injure others (Dellapenna, 1991).

In areas where water is scarce, trading in water quantity has occurred in the context of trading or selling one's right to capture and use waters of the state. An example is the Deschutes Water Exchange (DWE) in Oregon. The DWE Annual Water Leasing Program was established in 1998 as a cooperative effort with local irrigation districts and water-right holders to restore stream flows for environmental benefit. In that program, right holders are paid not to withdraw their rightful allocation of in-stream waters. The lease constitutes exercise of their water right so that it is not forfeited (in Oregon, a water right is forfeited if it is not exercised every five years), but the water that would otherwise be withdrawn is allowed to remain in stream (Deschutes Resources Conservancy, 2003).

However, trading in this case is to *increase* flows. We are not aware of any trading systems aimed at *reducing* the flow of water. Excess water is seen as a liability, not a desired asset to which rights and value attach. In this regard, water quantity trading to increase flows is fundamentally different from trading to decrease flows. With some important distinctions (discussed in a later section), trading to decrease stormwater flows is more analogous to pollutant trading where the value attaches to the right to discharge the item (i.e., pollutant), not to the item itself.

Implementation Scenarios

In the following sections, we present an interdisciplinary analysis of potential market-based instruments. Although these instruments can be employed in a variety of ways, for our purposes it is convenient to frame them in terms of price and quantity instruments. The instruments are examined from the perspective of hydrology, law, and economics. From the hydrology perspective, the instruments are compared on the basis of their relative potential to address the reduction of stormwater runoff. The legal analysis evaluates the alternative market-based instruments based on the legal complexity associated with implementation, which includes such considerations as (1) authority needed to implement the scenario,

(2) potential constitutional issues, (3) conflicts with existing law, (4) juris-dictional issues, and (5) other potential objections. Finally, the economics analysis focuses on the cost-effectiveness of the alternative approaches. Cost-effectiveness analysis identifies the instrument that can achieve a stated regulatory goal with the lowest compliance costs.

A few caveats need to be mentioned. Generalizing market-based instruments into the four price and quantity scenarios discussed below provides only a limited representation of potential market-based approaches. The hydrological, legal, and economic analyses are far from exhaustive, and do not consider all dimensions of the stormwater control problem. Instead, the purpose of this chapter is to discuss the opportunities along with some of the more onerous obstacles to creating market-based approaches for stormwater control. This focus on identi-fying the interdisciplinary complements and conflicts that may arise in implementing such options enables us to undertake this more narrowly defined analysis.

Price Instruments

Price instruments place a charge (either a fee or tax)* on the amount of stormwater runoff that a parcel generates, providing some incentive to reduce runoff to the point where the marginal control cost equals the runoff charge. To achieve the desired level of runoff control at the lowest cost, the stormwater runoff charge should be set equal to the marginal benefit from the desired level of reduced runoff. However, in practice this is not the case. In the following section the use of user fees and runoff charges are compared as price instruments for stormwater runoff control.

---

* The distinction between a fee and a tax varies from state to state, but in general, tax reve-nue can be used to fund legitimate government functions. The function need not be related to specific applications or purposes, but revenues are rarely earmarked for specific pur-poses and taxes must be applied uniformly to all taxpayers. In contrast, fees must bear a substantial relationship to the cost of providing the specified services and/or facilities and they must be apportioned in a fair and reasonable manner in order to be constitutionally valid. Traditionally, municipal stormwater systems were funded by tax revenue, but creat-ing fee-based stormwater management programs is becoming more popular. There are several likely reasons for this trend (Lindsey and Doll, 1999; Lehner. Apponte Clark, and Cameron, 1999): (1) not all local governments have authority granted by the state constitu-tion to collect taxes, but most, if not all can charge fees; (2) some state constitutions (such as Missouri and Michigan) require taxes (but not fees) to be subject to voter approval; (3) fees are perceived as more fair because the typical basis for determining the fee (e.g., an impervious surface) is an objective measurement, and because the fee reflects actual use of the system; (3) tax revenue is always subject to competing needs and changing priorities; and (4) unlike fees, taxes do not reflect the actual contribution of stormwater runoff.

Scenario #1: Stormwater User Fee

Stavins (2001) argues that most applications of price instruments have failed to have the incentive effects promised in theory, either because of the structure of the systems or because of the low levels at which charges have been set. Existing stormwater user fees may serve as an example of this phenomenon. As with any constitutionally valid fee, stormwater fees must be fair, equitable, and based on the cost of the service provided as measured directly or by some approximation of use or benefit. Stormwater fees are generally calculated by some measure of the approximate quantity of runoff that leaves a parcel of land, with the square footage of impervious surfaces being a broadly accepted metric. To provide incentives for the adoption of on-site source reduction BMPs, many stormwater utilities offer credits or fee reductions for landowners who implement BMPs to reduce runoff from their parcel. Table 8.B1 reproduced from Doll, Scodari, and Lindsey (1999) containing examples of these credits is given in Appendix B in this chapter.

Although user fees are directly related to excess stormwater runoff levels and have their revenues earmarked for the provision of closely related environmental services (i.e., construction or operation and maintenance of the stormwater sewer system), in practice they do not encourage desired behavior. Doll and Lindsey (1999) argue that existing user fee/credit systems, where available, have failed to encourage widespread adoption of on-site source reduction BMPs. From an economic standpoint, stormwater user fees are set too low to induce the appropriate level of BMP adoption on private property, and, therefore, are not a cost-effective means of stormwater runoff control.

There are several hydrological issues relating to the efficacy of using stormwater fees predicated on the extent of an impervious surface. Although setting an areal unit impervious surface equivalent to a certain volume of excess stormwater runoff is a step in the right direction, there are critical shortfalls in this method of estimation. Generalized notions of an impervious surface may be reasonable estimates of the two-dimensional extent of impervious surface, yet these values give no information on whether the impervious surface is connected to conveyances (i.e., connected impervious surface) to take stormwater away from the site, or otherwise allowed to flow onto the property. Furthermore, not all impervious surfaces are created alike. There are a very large number of combinations of impervious surfaces, which is complicated by differences in geometry (e.g., slope, curvature), that affect the timing and concentration of flows; and connectivity or proximity to pervious or other impervious surfaces, and distance to drains or other conveyances. These factors contribute to the production of tremendous variation in observed runoff volumes, and we speculate that these factors would become more important with a reduced parcel size. It seems that a more informed method of estimation would be required when considering runoff production in residential neighborhoods, or where land use patterns are relatively heterogeneous.

Additional hydrological concerns pertain to the use of stormwater user fees to influence appropriate BMP adoption. They include the limited range of sources subject to user fees, as well as the accuracy of the models used to predict stormwater contributions. Thus, landowners whose properties are served by a combined sanitary sewer system, as well as rural property owners, may be exempt from the user fee. This limits the potential range of stormwater runoff reduction and the areas that can be targeted for BMP adoption. In addition, fee reductions through credit provisions usually are limited to nonresidential properties, resulting in the exclusion of the expanding residential proportion of the total acreage within the watershed. These likely impair the ability of the stormwater fee system to reach the desired ecological target. The primary benefit of reliance on a stormwater user fee/credit system is that it has proven to be a legally acceptable price instrument in most jurisdictions.

Scenario #2: Stormwater Runoff Charge

To overcome the hydrological and economic shortcomings of existing stormwater user fee/credit systems addressed above, the stormwater utility needs to incorporate more accurate hydrological models, apply the price instrument to all landowners, and raise the price to reflect the marginal costs of reducing the desired level of runoff. This would convert the existing stormwater user fee into an incentive-based "stormwater runoff charge," where the charge would provide a more realistic "price signal" increasing its ability to influence private landowner behavior and coordinate that behavior across landowners.

To calculate an improved hydrologic status with a stormwater charge on runoff or impervious area, the hydrological model underlying the price mechanism must focus not only on the amount of impervious surfaces, but also an evaluation of landscape factors, extant development, existing stormwater and sewer infrastructure, and inventory of pipe breaks and other losses, and other information to understand better how the watershed drains. It is also important to incorporate the effect of spatial and transboundary relationships between adjacent parcels on stormwater runoff. For example, runoff may compound across many adjacent parcels, with implications for larger cumulative flows at the bottom of the slope. This is not a straightforward modeling effort. A model calibrated to local conditions and rainfall patterns can be parameterized to account for all of these factors, although routing these flows is beyond the scope of a lumped parameter model. A continuous, spatially distributed rainfall-runoff model could provide a reasonable approximation of the effect of stormwater runoff control on the hydrological status of the watershed. Model assessments can then calculate an average peak runoff from storm events within a 1- to 10-year return period. From the modeling exercise, one may identify conditions leading to the desired level of reduction in the average peak runoff.

From an economic standpoint, two sets of information are required for the implementation of a cost-effective stormwater runoff charge. First, the desired level of runoff reduction must be identified for a given recurrence interval. Second, the aggregate cost of stormwater runoff abatement from the landowners in the watershed needs to be determined. To influence behavior in a cost-effective manner the stormwater runoff charge must be set equal to the marginal aggregate cost of runoff control in the watershed. This provides the proper incentive for each landowner to provide the appropriate level of control to ensure the target is met. For this reason, price instruments are always difficult to implement effectively in practice. The costs of stormwater BMP installation, as well as the costs of operation and maintenance, are relatively well known. However, the private opportunity costs of individual landowners are not known and can be a substantial proportion of total costs. These costs include the foregone opportunity and use of the land that is devoted to BMPs. This information can be estimated, however, the penalty for not getting the price right can be substantial. When the price is set below the actual marginal cost, too little abatement will be provided. When the price is set above the actual marginal costs, the control standard is set too stringently. Although this problem can be overcome by starting with a reliable estimate of total costs, and revising the charge based on observed compliance, it could take a considerable amount of time to adjust to the appropriate charge level.

Legally, implementing a stormwater runoff charge poses considerable difficulty. If the price instrument were a tax, it would have to be applied uniformly (not based on actual use) and the proceeds would be available for any legitimate governmental purpose. In order to be a constitutionally valid fee, the charge must bear a substantial relationship to the cost of providing the specified services or facilities and it must be apportioned in a fair and reasonable manner. In some states, the cost of providing the service can include a consideration of related activities, allowing some increase in the fee such that it might provide incentives for changes in behavior, but the fee must still be substantially related to those activities. As an example, Ohio law* authorizes a local government to levy a fee on solid waste disposal for the purposes of defraying the added costs of maintaining roads, providing emergency services, paying for the costs of administering and enforcing the laws pertaining to solid wastes, and so on. Stavins (2001) reports that the unit

---

* Ohio Rev. Code Ann. §3734.57(C)(2003).

charge system used in the financing of household solid waste collection is most analogous to a Pigouvian tax. Furthermore, these incremental unit charges have been shown to reduce the generation of household waste (Stavins, 2001). Increasing the stormwater user fee to reflect the costs of retrofitting the stormwater sewer system to handle excess input could provide a solid incentive for BMP adoption. Actual implementation would depend on the specific laws of the state. In some states, user fees can fund only operation and maintenance and tax revenue is required for capital improvements. In others, fees are defined such that they can be used directly for capital improvements in the system. Other localities take an indirect approach by using the fee revenue to back loans or revenue bonds sold to finance the improvement (Keller, 2003).

The required connection between the fee and the cost of service and related activities applies to new development as well. Several states and localities now charge developers "impact fees" to cover the cost of new infrastructure and services needed to accommodate the added burden on the system. The rationale is that the current residents have already paid for the infrastructure that serves them (i.e., either through taxes or fees) and they should not have to pay for additional services to meet the new development's demands (Kolo and Dicker, 1993). Again, states vary in how they interpret and enforce these concepts and one could argue that current rate payers have not sufficiently "paid their way" if the existing infrastructure is not adequate to protect public health and the environment (i.e., as in the case of combined sewer systems).

Quantity Instruments

Quantity instruments can achieve the same cost-effective allocation of the control burden as a price system, with the added advantage of avoiding the problem of having to "get the price right." Under a quantity instrument (i.e., a trading system) an allowable overall level of stormwater runoff is established and allocated among individuals in the form of allowances. Landowners who keep their runoff levels below their allotted level may sell their surplus allowances to others. We explore two forms of quantity instrument. The first is a traditional allowance market, and the second is a modification that allows for the voluntary participation of individual landowners.

Scenario #3: "Cap-and-Trade" Stormwater Runoff Allowance Market

Allowance markets work in three stages: setting a cap, dividing the allowable runoff among parcel owners, and finally encouraging the trading of allowances. In terms of stormwater runoff, the cap is determined, if possible by a specific ecological constraint, such as stream flow capacity, or the cap might be determined by the capacity of the existing sewer system or other conveyances. The allowable runoff is then allocated among the individual sources in the market. These allowances grant permission to release a specified amount of runoff at the parcel level and could be based on the differential in stormwater runoff caused by development on the parcel. For example, any stormwater runoff from a 2-year storm event that would have occurred naturally (considering soil type, slope, and so on) when the property was in its predevelopment state would not have allowances associated with it. Allowances would only be needed for instances where runoff exceeds the predevelopment level (e.g., for Eastern states, this would be a forested condition). This system allows parcel owners to trade allowances, which creates incentives for individuals who can reduce runoff at lower costs to take on a larger share of the burden. For example, property owners could retain or detain stormwater, thus freeing allowances that could be traded with other property owners.

From the hydrological perspective, an allowance market is desirable because an enforceable limit is placed on the allowable runoff within the watershed. It also encompasses all current development within the watershed as well as any future development. Each landowner having either existing or new development would limit the postdevelopment runoff from his or her parcel either directly or through trading.

Economically, the allowance market overcomes the informational requirements of the price instrument, while promising the same level of cost-effectiveness. Trading can be accomplished through free-exchange, bilateral negotiation, or a clearinghouse arrangement. The choice of exchange mechanism is based on the tradeoff between administrative and transaction costs associated with each. The dispersed nature of stormwater runoff is difficult to observe at the individual parcel level. Resulting allowance markets, as in the case of existing water quality trading markets, may promote a monitoring exchange, based on the successful adoption and implementation of best management practices. To deal with the uncertainty of using an imperfect proxy for the actual abatement of these markets, trading ratios will also be incorporated.

In existing water quality trading markets, trading ratios have been introduced to allow for the exchange of heterogeneous goods within an allowance market. The trading ratio specifies the number of units of nonpoint pollution reduction, estimated by modeling the effectiveness of chosen best management practice, that must be exchanged for a single unit increase in point source pollution. A trading ratio greater than one provides a safety margin for the environment, as deviations from expected abatement performance of best management practices are less likely to result in violations of the regulatory standard. However, setting high trading ratios also reduces the benefits of trade and can lead to the collapse of the market altogether (Randall and Taylor, 2000). Stormwater allowance markets will face the same challenges. The ability of an allowance market to provide adequate protection of the overall quantity target, while also encouraging trading and its associated cost-savings, is an empirical question that must be determined on a case-by-case basis.

The stormwater runoff allowance market faces several legal challenges, particularly on properties with existing development. Property rights are viewed as "a bundle of rights." Examples include the drainage rights discussed above and the rights to possess, use, modify, lease, sell, or develop the land for residential, commercial, or industrial purposes (Daubenmire and Blaine, 2003). A state or its authorized political subdivision, such as a municipality may use its "police power" to restrict these rights if it is necessary to promote the public health, safety, and welfare; however, the power is limited by the Fifth and Fourteenth Amendments to the U.S. Constitution and also by many state constitutions. The Fifth Amendment prohibits the government from taking private property without just compensation. The Fourteenth Amendment prohibits states from taking property without due process of law. In this context, due process means the government regulation must have a legitimate goal and the method used to achieve that goal must be rationally related to the goal (Burgin, 1991). The prohibition on "takings" means the government cannot force "some people alone to bear public burdens which, in all fairness and justice, should be borne by the public as a whole" by taking their property without paying a fair price for it.*

Protecting the downstream environment from excessive flows could be considered a legitimate goal, but the question would be whether limiting a landowner's right to release diffused stormwater (by allowing less than current levels of runoff) is a rational way to meet that goal. In states using the "reasonable use" approach to drainage, the actions of the landowner, and nature of the harm could affect the answer. Where the "natural flow" rule is followed, hydrological parameters could factor into the decision. In states following a strict "common enemy" doctrine, a limit on stormwater releases might be viewed as an unacceptable means of achieving the goal.

Whether the property is currently developed or undeveloped could also affect the outcome. Several options exist for limiting stormwater runoff from new development or redevelopment including limiting the amount of impervious surface permitted on a property through zoning, encouraging or requiring the use of low-impact designs, and installing BMPs as part of the development (Livingston et al., 1997). Another approach is to avoid development altogether by purchasing or transferring the development rights associated with the property (Daubenmire and Blaine, 2003).

---

* Armstrong v. United States, 364 U.S. 40, 49 (1960).

Limiting stormwater runoff from currently developed properties may be more problematic. Aside from possible political objections, it might be challenged as a retroactive law that unconstitutionally changes the legal status of a vested right. Ohio's constitution, for example, prohibits the passage of laws that affect substantive rights or impose new or additional burdens, duties, obligations, or liabilities as to a past transaction.* Depending on the nature of the limit, a property owner could argue that it interfered with the development rights that were vested in the property at the time the improvements were made. On the other hand, if a strong enough case can be made for the need to limit runoff to protect human health and welfare, the restriction on existing development might be seen as a legitimate exercise of the police power. In that regard, any potential nuisance or health issues associated with the BMP would have to be considered as well.

If a regulation constitutes a permanent physical occupation of land without compensation (e.g., using private property for a public drainway without paying for it), it would be a taking regardless of the interference with the owner's use or the importance of the government's interest.† In any event, the specific facts of each case would have to be examined by a court to determine if the regulation were valid or if a taking occurred.

Scenario #4 Voluntary Offset Program

A voluntary offset program using the existing NPDES permit program as an incentive for government participation would provide economic incentives for private landowners to reduce their stormwater runoff. Publicly owned treatment works (POTWs) and municipal separate storm sewer systems (MS4s) are both subject to NPDES permits. In the case of POTWs, the permits set effluent limitations and treatment requirements. Stormwater runoff plays a contributing factor in POTW control in the presence of combined sewer systems. In the presence of CSSs, a POTW will have the incentive to purchase BMPs that reduce runoff from landowners. These reductions avoid the need for more costly sewer system upgrades by reducing the incidence of CSS overflows. For MS4s, their permits simply require that they limit the amount of pollutants entering the sewer system. In some cases, this may be accomplished at lowest cost by reducing the amount of sediment and pollutant-bearing runoff entering the system at the parcel level. This trading program allows regulated POTWs and MS4s to meet their existing regulatory obligations by financing the adoption of BMPs by private landowners. This scenario is analogous to the Deschutes Water Exchange in which the government pays the

---

* Ohio Const. §28, art. II. (2003).
† Loretto v. Teleprompter Manhattan CATV Corp., 458 U.S. 419 (1982).

water-right holder to not exercise his or her right to withdraw water from the stream. In this case, the government is paying the landowner to not exercise her right to release all of her stormwater to the POTW or MS4.

MS4s do not have as stringent regulatory obligations as do POTWs because they are not subject to particular effluent standards or treatment standards.* MS4s may be able to easily meet their NPDES requirements without the need to reduce the amount of runoff entering their system. MS4s are not regulated on the amount of runoff in their systems but on their practices of limiting pollutants from entering stormwater runoff. MS4s may not find a great benefit in this type of trading scenario. POTWs with CSSs, though, would greatly benefit from the reduction of stormwater runoff entering their system because the amount of stormwater is the cause of the combined sewer overflows during storm events (USEPA, 2004).

From a legal perspective, because the offset program incorporates landowner participation on a voluntary basis, many of the constitutional issues presented in an allowance trading market are avoided. One issue that needs to be considered is that revenue, be it from taxes or fees, must be applied to a legitimate government purpose. Thus, paying landowners to reduce their stormwater runoff would have to be considered a legitimate government purpose for constitutional purposes. This should not be difficult to demonstrate, particularly in the case of the POTW with a CSS inasmuch as a reduction in flows would address a significant public health issue. However, the applicability of this type of trading scenario is most likely limited to landowners within the reaches of a POTW, MS4, or both systems. A POTW or MS4 may be limited to using fee revenue to benefit only those who have contributed to these funds, that is, landowners within the system. Few limitations would apply to tax revenue (but the local authority may not have the power to tax). If the POTW or MS4 or some other governmental entity has the authority to pay landowners outside the utility system, then the system could benefit from reductions in stormwater runoff at other locations within the watershed.

This program would be most accurately described as a combination price/quantity mechanism. If the POTW or the MS4 receives a substantial portion of their operating budget, including the funds to pay for landowner BMP adoption, through tax revenue that is collected by the municipality

---

* Although an MS4 also has an NPDES permit for releases of pollutants introduced into the system, its requirements are limited to implementing six control measures under its NPDES permit: (1) public education and outreach on stormwater impacts; (2) public involvement/participation; (3) illicit discharge detection and elimination; (4) construction site stormwater runoff control; (5) postconstruction stormwater management in new development and redevelopment; and (6) pollution prevention/good housekeeping for municipal operations. National Pollutant Discharge Elimination System—Regulations for Revision of the Water Pollution Control Program Addressing Storm Water Discharges, 64 Fed. Reg. 68722, 68736 (Dec. 8, 1999).

and allocated to the utility, the adoption of this offset program (i.e., quantity instrument) could result in increased taxes (i.e., a price instrument). However, for simplicity we are assuming that the tax pool is large enough that the increase in individual tax payments to pay for the offset program is relatively small and will not have any effect on individual behavior. It is important to note that in the case where the increase in the tax payments of landowners is noticeably affected by the adoption of the offset program, important synergies between the two would have to be examined.

From the perspective of economic theory, the switch to voluntary participation should not change the outcome of the quantity instrument, as long as transaction costs are low and the full right to release or the obligation to detain stormwater is exchanged. However, it is not likely that either of these assumptions holds true in implementation. The utility would enter into a private agreement with each landowner specifying the amount of compensation for the amount of stormwater reduction, but the liability of compliance with the CWA will remain with the utility. Recourse for breach of contract would have to be sought through the court system, adding the costs of verifiable monitoring and enforcement of BMP adoption and performance to the POTW or MS4s transaction costs.

From a hydrological standpoint, the voluntary offset program is preferable to the price instruments because it identifies an ambient quality standard that must be met, and it avoids the uncertainty of setting the appropriate price. Yet this does not directly pertain to controls on the quantity of stormwater or its direct effects on quality. This implementation may not be appropriate for smaller municipalities, based on the lack of connected septic sewers, absence of an even moderately sized POTW, and the variability in the specific arrangement of hydraulic conveyances.

This implementation also is inferior to the allowance market, because the "cap and trade" component is lost. Landowners retain the right to unlimited release of stormwater runoff from their parcel. Therefore, changes to impervious surfaces are neither prevented nor discouraged through this program, and this threatens the ability of maintaining the overall standard through this type of trading arrangement. The only option for maintaining the overall standard, in the presence of increased impervious surfaces on existing properties, is to require the POTW or MS4 to control an increasing amount of stormwater runoff through their NPDES permits.

Synthesis and Conclusion

As expected, no clear cross-disciplinary consensus emerges in the choice of a market-based approach to control stormwater runoff. The selection of an appropriate management instrument requires complicated tradeoffs between the important goals of all three disciplines and their preferred implementation scenarios. In practice, these tradeoffs are dependent upon the unique physical characteristics of the watershed, as well as the existing legal structure and social institutions of the community. Table 8.1 provides a brief summary of the disciplinary concerns associated with each of the scenarios discussed above.

TABLE 8.1

	PRICE INSTRUMENTS		QUANTITY	
	Scenario #1 Stormwater User Fee	Scenario #2 Stormwater Runoff Charge	Scenario #3 Allowance Market	Scenario #4 Voluntary Offset Program
Economic Concerns	Improperly priced incentive results in too little investment in BMPs	Cost-effective when fee is set equal to marginal cost of desired runoff reduction; Extensive private cost information required	Cost-effective; Reduced information requirement	Voluntary participation by landowners reduces incentive effects
Ranking	4	2 or 3	1	2 or 3
Hydrologic Concerns	May not include all parcels; error-prone accounting for impervious surface	May not include all parcels; complex modeling effort required with at least some initial monitoring	Includes all parcels; difficult to determine a "cap" on allowable total runoff; monitoring required	May not include all parcels; No "cap" on impervious surface or total allowable runoff; quality versus quantity issues; monitoring required
Ranking	4	2 or 3	1	2 or 3
Legal Concerns	Program currently exists	No constitutional authority to charge incentive-based fees	Constitutional issues, particularly on existing development	Voluntary; authority may currently exist
Ranking	1	4	3	2

Providing even a simple rank order based on the disciplinary criteria for each of the scenarios proves to be difficult. We provide a general ranking, while acknowledging that the ordering will not hold in all cases. As mentioned above, the hydrological rankings are based on their potential to address environmental quality goals, the economic rankings are based on the cost-effectiveness of the alternative approaches, and the legal rankings are based on the legal complexity associated with implementation. These various factors contribute to the difficulty in making definite rankings and therefore some rankings are somewhat interchangeable. Moreover, no clear choice of implementation scenario is apparent from these rankings. This shows that not one scenario is the solution to the stormwater runoff problem.

We applied this interdisciplinary approach to the stormwater runoff problem to create several stormwater management scenarios. Based on these rankings, our next step is to pursue further research on a voluntary offset program. Although it is not the first choice of any discipline, it best addresses the concerns of all the disciplines involved. The voluntary offset program will prove to be a challenging application of this interdisciplinary approach and it is hoped will provide beneficial results for state and local governments to apply to the reduction of stormwater runoff in their watersheds.

Acknowledgments

This chapter is reprinted from *Environmental Science and Policy*, Vol. 8/Issue 2, Punam Parikh, Mike Taylor, Theresa Hoagland, Hale W. Thurston, and William Shuster. At the Intersection of Hydrology, Economics, and Law: Application of Market Mechanisms and Incentives to Reduce Stormwater Runoff, pp. 133–144. Copyright 2005, with permission from Elsevier

The views expressed herein are strictly the opinions of the authors and in no manner represent or reflect current or planned policy by the USEPA. This research was performed while Punam Parikh held both a National Research Council Associateship Award and a Federal Postdoctoral position and while Michael A. Taylor held a National Research Council Research Associateship Award at the United States Environmental Protection Agency National Risk Management Research Laboratory. The authors also wish to thank Audrey Mayer and Haynes Goddard for helpful feedback on earlier drafts of this manuscript.

Appendix A

The marginal abatement costs for two individuals are illustrated in Figure 8.A1. The marginal cost of abatement represents the additional outlay that is required to reduce one additional unit of output (i.e., run-off). The area under the marginal abatement curve gives the total costs of abatement. Thus, in Figure 8.A1, Mr. A has lower abatement costs relative to Ms. B. Suppose that the regulator desires a total of 14 abatement units from these individuals. Under a command-and-control approach both Mr. A and Ms. B are given uniform standards requiring each to abate seven units. As shown in Figure 8.A1, this will result in different marginal costs of abatement and, therefore, is not the lowest cost allocation.

Again, we assume the regulator is interested in achieving 14 units of total abatement, but this time at the lowest possible abatement costs. Figure 8.A2 illustrates how a price-based instrument ensures the lowest cost allocation of the abatement burden. Faced with a per-unit charge of T, both Mr. A and Ms. B will choose to provide some level of abatement. In particular, they both will choose to provide abatement as long as it is less costly than the per-unit tax. In Figure 8.A2, setting the tax at price T will result in Mr. A choosing to abate 9 units and Ms. B choosing to abate 5 units, for a total of 14 units of abatement. It is important to note that setting the tax at the appropriate price is essential to achieving the desired level of

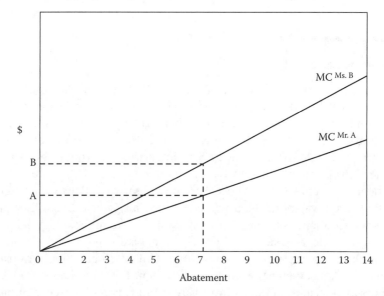

FIGURE 8.A1
Command-and-control.

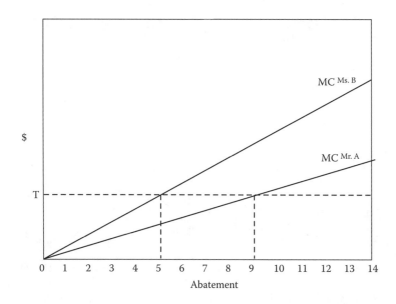

FIGURE 8.A2
Price instrument (Pigouvian tax).

total abatement. It is easy to see in Figure 8.A2, if T were set higher than its current level each individual would provide more than the required level of abatement. This results in unnecessary abatement costs. Similarly, if T were set lower than in Figure 8.A2, each individual would provide less abatement resulting in less than the desired level of total reductions. This requires either the regulator to have detailed information regarding individual abatement costs, or the use of an iterative system of tax setting. The iterative process updates the tax level at the end of each year, based on the observed level of total abatement.

Figure 8.A3 illustrates an allowance market where the desired level of total abatement is 14 units. Assume the control agency has initially distributed allowances to Mr. A and Ms. B such that each is required to abate 7 units of runoff. Note that this is the same allocation as under the command-and-control example in Figure 8.A1. However, both individuals have an incentive to reallocate the abatement responsibility. Ms. B will be willing to pay any price less than P^B for each unit of abatement that Mr. A is willing to provide, because this will reduce her total cost of abatement. Mr. A is willing to provide additional abatement at any price greater than P^A, as this will provide him with profit. The equilibrium price will be P^*, where the marginal costs of abatement of Mr. A and Ms. B are equal and no further benefits of trade exist. This also results in the low cost allocation of the abatement, where Mr. A abates 9 units of runoff and Ms. B abates 5 units.

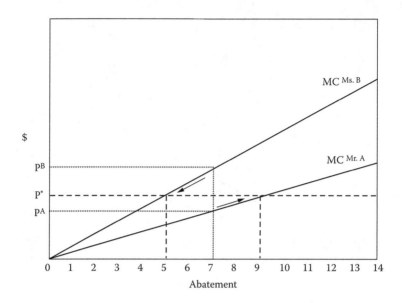

FIGURE 8.A3
Quantity Instrument (Allowance Market).

Appendix B
TABLE 8.B1

Summary of Credit Options[a]

Utility	Eligible Users	Basis for Credit	Design Storm	Maximum Credit	Typical Credit
Gainesville, FL	Nonresidential properties	Volume of on-site detention	25-year, 24-hour	100% of base fee	15–35%
Orlando, FL	Commercial and multifamily residential	On-site retention or detention	NA	42%	42%
Wichita, KS	Properties less than or equal to 50 ERUs	Two credits: volume of detention or retention	1. 100-year 2. Complete retention	1. 40% 2. 80%	Currently no applications
Louisville–Jefferson County KY	Commercial properties	On-site detention of peak flows	2-year, 10-year and 100-year predevelopment runoff	82%	Varies with degree of control
St. Paul, MN	Nonresidential properties	On-site detention of peak flows; acreage, peak flows	5-year and 100-year release limited to (1.64 ft^3/ac/s)	10% (5-year) 25% (100-year)	Varies with degree of control
Charlotte, NC	Commercial, industrial, institutional, multifamily, residential, and homeowner associations	1. Peak discharge 2. Total runoff volume 3. Annual pollutant loading reduction	1. 10-year, 6-hour 2. 2-year, 6-hour 3. Reduction in loading	1. 50% 2. 25% 3. 25% Up to 100%	Varies with degree of control

continued

TABLE 8.B1 (Continued)

Summary of Credit Options[a]

Utility	Eligible Users	Basis for Credit	Design Storm	Maximum Credit	Typical Credit
Durham, NC	Nonresidential properties	Pollution credits for water quality and quantity controls	State standards for facility design; estimated pollutant runoff efficiency	25%	Few applications received
Cincinnati, OH	Commercial properties	On-site retention	Limited discharge to predevelopment runoff	50%	Credit never used
Tulsa, OK	Privately maintained facilities	50% or greater detention; maintenance costs of on-site facilities	NA	60%	Varies
Austin, TX	Commercial properties	On-site detention; inspection	NA	50%	50%
Bellevue, WA	All properties	On-site detention; intensity of development	NA	Reduction of one rate (intensity of development) class	Varies
King County, WA	Commercial properties	Private maintenance	NA	Reduction of one rate class	Varies
Indianapolis, IN	Nonresidential properties	Discharge to specified streams; on-site retention or detention watershed size	Tier Two: 2-, 10-, 25-, 50-, 100-year events	Tier One: 24%; ≤$50 Tier Two: 35%; <$250	(Proposed)

[a] Doll, A., P.F. Scodari, and G. Lindsey, 1999. *Credits as Economic Incentives for On-Site Stormwater Management: Issues and Examples.* Proceedings, U.S. Environmental Protection Agency, National Conference on Retrofit Opportunities for Water Resource Protection in Urban Environments, Chicago, IL, February 9-12, 1998, EPA/625/R-99/002, July 1999, pp. 13-117.

References

Booth, D.B., and C.R. Jackson. (1997). Urbanization of aquatic systems – Degradation thresholds, stormwater detention, and limits of mitigation. *J. Amer. Water Resources Assoc.*, 33: 1077–1090.

Burgin, S. (1991). Local governments taking charge of water quality—Is it a good idea? *Natural Resources Environ.*, 5(4), American Bar Association, Chicago, IL pp. 20–22.

Cairns, J. (1995). Urban runoff in an integrated landscape context. In E.E. Herrick (Ed.), *Stormwater Runoff and Receiving Systems Impact, Monitoring, and Assessment.* Boca Raton, FL: CRC Press, pp. 9–20.

Carlson, T.N., and S.T. Arthur. (2000). The impact of land use – Land cover changes due to urbanization on surface microclimate and hydrology: A satellite perspective. *Global Planetary Change*, 25(1–2): 49-65.

Church, P.E., G.E. Granato, and D.W. Owens. (1999). Basic Requirements for Collecting, Documenting, and Reporting Precipitation and Stormwater-Flow Measurements. *USGS National Highway Runoff Data and Methodology Synthesis.* Northborough, MA.

Daubenmire, J., and T.W. Blaine. (2003). Purchase of Development Rights, CDFS-1263-98, Ohio State University, Columbus OH pp. 1-5. URL: http://www.ohioline.osu.edu/cd-fact/1263.html (last visited, 11/2003).

Dellapenna, J.W. (1991). The legal regulation of diffused surface water, *Villanova Environ. Law J.* (2 Vill. Envtl. L.J. 285), pp. 285–332.

Deschutes Resources Conservancy, The (2003). Deschutes Water Exchange, http://www.deschutesrc.org (last visited Dec. 22, 2003).

Doll, A., and G. Lindsey. (1999). Credits bring economic incentives for onsite stormwater management. *Watershed Wet Weather Tech. Bull.*, 4(1): 12–15.

Doll, A., P.F. Scodari, and G. Lindsey. (1999). Credits as economic incentives for on-site stormwater management: Issues and examples. In *Proceedings, U.S. Environmental Protection Agency National Conference on Retrofit Opportunities for Water Resource Protection in Urban Environments,* Chicago, February 9–12, 1998, EPA/625/R-99/002, July 1999, pp. 13–117.

Environomics. (1999). A Summary of U.S. Effluent Trading and Offset Projects, prepared for Dr. Mahesh Podar, U.S. Environmental Protection Agency Office of Water.

Green Nature (2002). This cloud has no silver lining: Acid rain in the Northeast, http://greennature.com/article942.html.

Hey, D.L. (2001). Modern drainage design: The pros, the cons, and the future. In *Hydrologic Science: Challenges for the 21st Century.* Bloomington, MN: American Institute of Hydrology.

Jauregui, E., and E. Romales. (1996). Urban effects on convective precipitation in Mexico City. *Atmospher. Environ.*, 30(20): 3383–3389.

Johnson, S.L., and D.M. Sayre. (1973). Effects of urbanization on floods in the Houston, Texas metropolitan area. United States Geological Survey Water Resources Inventory, 3–73, 50 p.

Keller, B.D. (2003). Buddy can you spare a dime? What's stormwater funding, *Stormwater*, 2(2): 38–72.

Klein, R.D. (1979). Urbanization and stream quality impairment. *Water Res. Bull.*, 15(4): 948–963.

Kolo, J., and T.J. Dicker. (1993). Practical issues in adopting local impact fees. *State Local Govern. Rev.*, 25(3): 197–206.

Lehner, P., G.P. Apponte Clark, and D.M. Cameron. (1999). Stormwater Strategies: Community Responses to Runoff Pollution, Chapter 4, Natural Resources Defense Fund, 145 pp. Available at URL: http://www.nrdc.org/water/pollution/storm/chap4.asp

Lindsey, G., and A. Doll. (1999). Financing retrofit projects: The role of stormwater utilities. In *Proceedings of the National Conference on Retrofit Opportunities for Water Resource Protection in Urban Environments*, Chicago, IL, February 9–12, 1998, EPA/625/R-99/002, July 1999, pp. 107–112.

Livingston, E.H., E. Shaver, R.R. Horner, and J.J. Skjpien. (1997). *Institutional Aspects of Urban Runoff Management: A Guide for Program Development and Implementation*. Ingleside, MD: Watershed Management Institute.

National Pollutant Discharge Elimination System. (1999). Regulations for Revision of the Water Pollution Control Program Addressing Storm Water Discharges; Final Rule, 64 Fed. Reg. 68,722 (Dec. 8).

Neller, R.J. (1988). A comparison of channel erosion in small urban and rural catchments, Armidale, New South Wales. *Earth Surface Process. Landforms*, 13: 1–7.

New York State Department of Environmental Conservation. (2004). Acid Rain, http://www.dec.state.ny.us/website/dar/ood/acidrain.html

Putnam, A.L. (1972). Effect of urban development on floods in the Piedmont Province of North Carolina. United States Geological Survey open-file report, 87 p.

Randall, A., and M.A. Taylor. (2000). Incentive-based solutions to agricultural environmental problems: recent developments in theory and practice. *J. Agric. Appl. Econ.*, 32, 2(August): 221–234.

Stavins, R.N. (2001). Experience with Market-Based Environmental Policy Instruments Resources for the Future Discussion Paper 01-58.

Swanson, F.J., S.L. Johnson, S.V. Gregory, and S.A. Acker (1998). Flood disturbance in a forested mountain landscape. *BioScience*, 48(9): 681–689.

Tomasi, T., K. Segerson, and J. Braden. (1994). Issues in the design of incentive schemes for nonpoint source pollution control. In *Nonpoint Source Pollution Regulation: Issues and Analysis*. Dordrecht: Kluwer Academic. pp. 1–37.

USEPA. (2004). August 26. Report to Congress on Impacts and Control of Combined Sewer Overflows and Sanitary Sewer Overflows. EPA 833-R-04-001. http://cfpub.epa.gov/npdes/cso/cpolicy_report2004.cfm pp. 2–3.

U.S. GAO. March (2000). Report to Congressional Requesters: Acid Rain, Emissions Trends and Effects in the Eastern United States. GAO/RCED-00-47.

Walesh, S.G. (1989). *Urban Surface Water Management*. New York: Wiley-Interscience

9

In-Lieu Fees: Steps Toward Stormwater Treatment Cost-Effectiveness

Chelsea Hodge and W. Bowman Cutter

CONTENTS

Introduction

As has been noted elsewhere in this book, regulations increasingly require cities of all sizes to manage stormwater runoff so that the speed, quantity, and quality of the runoff do not adversely affect receiving water bodies. As a result, city stormwater responsibilities have moved beyond the traditional emphasis on flood control to water quality concerns. Cities use both regional and on-site strategies to meet these new responsibilities. Regional stormwater control concentrates runoff from a large area and treats the collected stormwater at a centralized facility on public land. Besides constructing stormwater best management practices (BMPs) on public land, many municipalities enact a stormwater policy requiring developers of new large commercial and industrial properties to install on-site BMPs. Requiring or incentivizing developers and landowners to manage runoff on site decreases the need for regional runoff management and reduces local government runoff management costs.

However, on-site BMPs can be costly to developers, difficult to construct on some properties, and difficult to maintain. Difficult site geology or shape can result in high costs. If improperly sited or poorly constructed, placing BMPs on numerous small sites could result in many slightly useful but difficult to maintain BMPs. It can be politically difficult to impose costs on the private sector for what some see as a government responsibility.

A small but growing number of municipal stormwater management policies contain an in-lieu fee (ILF) that mitigates the economic and political costs of on-site BMP requirements. On-site stormwater management policies containing an ILF allow developers and property owners to pay a fee in lieu of meeting all or a portion of the regulatory requirement. The local government uses the collected funds to provide runoff management elsewhere in the community. ILFs have the potential to provide the predictability of a traditional direct regulation while allowing some of the flexibility and economic efficiency associated with purely incentive-based policies.

Despite their potential, stormwater ILFs have received little attention in the published literature. The only existing literature on stormwater quantity ILFs are explanations of the policy that can be found in sample ordinances and manuals for stormwater professionals. We have not located past research that catalogs or surveys existing policies, presents a list of best practices, compares alternative pricing structures, or otherwise presents a meaningful analysis of ILFs.

This chapter presents a survey of approximately 30 municipalities whose stormwater management policy includes an ILF. Our goal was to answer two questions:

1. Why do cities adopt ILFs as part of their stormwater programs?
2. Are ILFs structured to improve the cost-effectiveness of the stormwater management system to the best of their ability?

We begin with an introduction to the stormwater runoff regulatory landscape. Then we discuss previous literature on ILF options in stormwater policies. Next, we discuss the theoretical advantages of an in-lieu system. Finally, we present the results of our survey and conclude.

Regulatory Requirements and On-Site BMPs

It has been noted throughout this volume that regulations on urban stormwater runoff have gradually been strengthened since the late 1980s. The National Pollutant Discharge Elimination System (NPDES), a part of the Clean Water Act of 1977, was originally construed to regulate point sources of pollution such as factories and publicly operated treatment works (POTWs). Since 1977, however, NPDES Phase I and II regulations have extended responsibility for managing stormwater runoff to first large and then small municipalities. Under NPDES regulations, cities and other local bodies are treated as point-source polluters with the responsibility to meet water quality standards in their runoff discharges to stormwater systems or water bodies (Moffa 1996).

Cities' urban stormwater management duties grew further with the introduction of total maximum daily load (TMDL) legislation Whereas NPDES regulations are source-based regulations, in which compliance is measured by the quality of the discharge, TMDL legislation is a type of receiving water body regulation, in which compliance is determined by whether the aggregate discharges to receiving water bodies endanger the beneficial uses of the water body, such as water supply, fishing, swimming, and so on. The result of TMDL regulation is to tighten NPDES regulatory requirements significantly, because although urban runoff may not directly violate water quality standards, it could well be a contributing factor in degrading a river, bay, or lake. In addition, the water quality emphasis in the NPDES and TMDL regulation means that cities must control and manage runoff from both large storms, the historic focus of stormwater management policies, as well as small storms.

These regulations, as well as state and local water quality concerns, have caused most counties and cities to implement a set of stormwater regulations to improve the quality of water discharged into their water bodies, mitigate the deleterious effects that high peak water flows can have on streams and rivers, and reduce the chances of flooding. A key component of most of these regulations is requirements for on-site stormwater management. Although the definition of "on-site stormwater management" varies from one municipality to the next, these regulations typically require new development and redevelopment meeting certain criteria to construct on-site devices that detain (occasionally retain) stormwater or remove pollutants. Stormwater detention is simply the act of building a basin, lake, culvert, or other stormwater management device to collect and detain stormwater flowing off a site's roofs, parking lots, and other hardscape. These devices release stormwater into the municipal stormwater conveyance system at a rate much lower than would be observed if the device were not in place. Detention or retention can also partially treat the runoff by slowing down the water, which allows particulates and some pollutants to settle out before the water moves off-site (Ferguson, 1998). Some stormwater approaches, such as rain gardens, infiltrate to groundwater and remove pollutants through soil filtration.

The property classes required to install on-site detention devices vary from municipality to municipality. Most municipalities that have an on-site stormwater quantity management regulation require that devices be installed during new construction of large commercial and industrial developments (Jenkins and Herricks, 1995). Existing development and residential areas are typically excluded from on-site stormwater management requirements, however, some municipalities require new residential subdivisions to construct an on-site stormwater management system. A smaller number of municipalities require installation of stormwater management devices when a property is being redeveloped (Jenkins and Herricks, 1995). However, because on-site systems installed at new construction sites only treat a portion of the runoff, most cities must construct regional stormwater facilities to treat stormwater from developed properties in order to comply with NPDES and TMDL regulations. A challenge facing cities in their new water quality responsibilities is balancing on-site management with regional detention.

Cost-Effectiveness of On-Site Placement

On-site stormwater management varies significantly in its cost-effectiveness from property to property (Cutter et al., 2008). In other words, in some cases on-site stormwater management can manage runoff at a lower cost per acre-foot of runoff managed than centralized regional management, whereas in other cases, the cost of on-site management is comparable to or higher than the cost of regional management. To ensure that on-site BMPs installed under a policy are cost-effective, the following must be true: (1) installed on-site management must be less expensive than equivalent incremental additions to regional runoff management, and (2) on-site requirements must be structured so that the incremental cost of the added runoff management capacity is approximately equal across sites.

The costs of on-site BMPs include capital, labor, maintenance, and land. Different soils and site topography across sites that result in dissimilar labor and capital costs for the same runoff capture effectiveness. Even similarly sized sites can produce different amounts of runoff depending on their imperviousness and soil type. These and other factors influence the cost of stormwater BMPs and the quantity of runoff that they must manage.

Land value is likely the main driver of differences in cost-effectiveness across sites. Land values can differ radically between dense urban center areas and outlying suburbs (Dye and McMillen, 2007). In addition, the value of the land to the owner can differ even across properties in the same vicinity. For example, some properties may have landscaping that allows a stormwater BMP to be incorporated at little or no added land cost, thus making the land portion of the total cost of the BMP very small.

Because of the variation in on-site BMP cost-effectiveness, uniform regulatory requirements impose high costs per volume of stormwater managed on some landowners and low costs on others. Basic economic principles dictate that on-site management should be expanded where it is cost-effective and reduced where it is not.

As listed above, our first criterion for cost-effective policy is that the policy should ensure that the marginal cost of newly installed on-site BMPs be less than the marginal costs of expanding regional treatment. (By marginal cost, we mean the cost of a small increase in the capacity to manage runoff.) Cost-effective stormwater regulation should seek to place lower-capacity BMPs on particularly expensive parcels, especially where the marginal costs of those BMPs is greater than the alternative of expanding regional treatment.

It is key to consider marginal (incremental) rather than average cost in this comparison. The average cost for regional treatment could be low and the marginal costs be high. Consider a stormwater infiltration basin that is to be placed on a large parcel of public land and is already sized to occupy the public parcel fully. This is likely to be a low average-cost system because the land is already owned by a public entity and the large size of the basin implies there are significant economies of scale. However, expansion of regional treatment could entail acquiring additional land or placing basins on smaller properties with lower economies of scale. The incremental costs of regional stormwater management could be significantly larger than average, and on-site BMPs that appear expensive relative to the average cost of regional stormwater management would be good investments.

The second criterion for ensuring that a stormwater policy is cost-effective is that, on the margin, the cost-effectiveness of new on-site BMPs should be equal. In other words, on-site BMPs should be sited so that an extra dollar spent on slightly greater capacity at any facility captures the same amount of added runoff.* In practice, this would mean that cost-effective siting of on-site BMPs would result in the installation of more capacity on properties that have good soils, low land values for their area, or are otherwise inexpensive. Properties with high on-site BMP costs would either not install a BMP or would install a BMP with a smaller capacity.

Economists have proposed various policies to increase the cost-effectiveness of stormwater management solutions, such as runoff taxes and cap and trade systems (see Thurston 2006; Thurston et al., 2003,

---

* Capacity can be defined many ways in stormwater management. The definition here is void capacity in an infiltration or detention system.

2008; and Cutter et al., 2008). ILFs are another policy instrument that can move stormwater regulation toward cost-effectiveness. ILFs allow properties where it is difficult or costly to place BMPs to pay a fee in return for an exemption from BMP requirements or a downscaling of the required BMP size.*

ILFs can be structured to meet the first cost-effectiveness criterion. To do so, fees should be set approximately equal to the marginal cost of regional treatment expansion that would be needed if on-site BMPs are not constructed. Because these marginal costs are likely to change over time, in-lieu fees should be flexible and determined by a formula that considers inflation and changes in costs over time. ILFs cannot be structured to meet the second criteria of equaling cost-effectiveness across on-site BMPs because they do not provide an incentive for developers or property owners to go beyond normal development requirements for sites where there are inexpensive options for additional stormwater management.

Literature on In-Lieu Fees

Despite their potential, stormwater ILFs (and ILFs used in other applications) have received little to no attention in the published literature. The only existing literature on stormwater quantity in-lieu fees are explanations of the policy that can be found in sample ordinances and manuals for stormwater professionals. There is little past research that catalogs or surveys existing policies, presents a list of best practices, compares alternative pricing structures, or otherwise presents an analysis of in-lieu fees.

Jenkins and Herricks (1995) state that most municipalities require all developments to build detention facilities so developers face similar costs. A "fee-in-lieu of construction" policy is seen as a way to ensure that all developers face similar development costs while not building unnecessary structures that do not contribute to peak-flow runoff control (p. 326). However, equalizing cost is not the same as equalizing cost-effectiveness. Cost-equalization calls for installing all BMPs with similar costs even when one location offers a better stormwater return on the dollar because of better soils, less expensive land, or other factors. Cost-effectiveness will place greater BMP capacity on good locations and less capacity on poor locations.

---

* From an economics point of view, the difficulty of BMP placement is synonymous with costly placement: a developer or property owner can always handle stormwater regulatory requirements by developing the property with a smaller footprint or otherwise altering the property so that BMPs can be fit into development plans. However, it can be extremely expensive to do so because reducing the size of the building footprint could significantly reduce the overall value of the property.

The Model Stormwater Drainage and Detention Ordinance published by the Northeastern Illinois Planning Commission in 1990 argues for in-lieu fees essentially on cost-effectiveness grounds. The ordinance states that requirements for BMP placement on all newly developed sites can result in numerous, small BMPs that are difficult to monitor or maintain. In-lieu fees can weed out the smaller, less cost-effective on-site BMPs. The guide provides a list of tasks that should be completed by a municipality before implementing a fee in lieu of a detention program.

The Maryland Department of the Environment (2008) discusses the legal status and financing impacts of ILFs. Debo and Reese (2003) discuss in-lieu fees as a funding mechanism. They both note that ILFs are a promising source of revenue to fund regional solutions. This is consistent with cost-effectiveness economic arguments: where on-site BMPs are of a higher cost than incremental additions to regional solutions, then the ILF is more valuable than the on-site BMP for reaching regional stormwater management goals.

Several sources (Nelson, 1995; Center for Watershed Protection, 2008; Virginia Department of Conservation and Recreation, 2001) discuss different approaches to calculating stormwater in-lieu fees. Nelson (1995) recommends that in-lieu fees be calculated based on a local hydrological study to estimate the cost of constructing alternative stormwater storage facilities. Nelson (1995) also recommends that if in-lieu fees are charged per unit (e.g., square foot) of development, fees should vary based on the zoning district to account for the variations in typical imperviousness of different types of development. Various ordinances and development codes have similar recommendations for charging in-lieu fees based on the volume of runoff that would be managed by the on-site system (e.g., Virginia Department of Conservation and Recreation, 2001; Center for Watershed Protection, 2008).

Charging in-lieu fees based on the volume of runoff (or possibly peak runoff for a criteria storm) that would have been detained/retained by the on-site BMP is likely a more cost-effective basis for in-lieu fees than charging based on the capacity not installed. This is because similar size properties could generate different amounts of runoff depending on the soil type and impervious area.

Stormwater utilities sometimes include credits allowing landowners that implement on-site BMPs to reduce or eliminate their stormwater utility fee. Stormwater credits are economically similar to in-lieu fees. Landowners with low costs of implementing BMPs will choose to install them rather than pay the fee. For cost-effectiveness, the credit should be set low enough so the on-site BMP contributes more to runoff management than the stormwater utility fees would have contributed to regional solutions. Credits can improve cost-effectiveness by providing incentives for particularly cost-effective BMP placements. In-lieu fees increase the overall cost-effectiveness of the system by weeding out non-cost-effective on-site BMP placements. Doll et al. (1998) survey various cities with credit programs on the eligibility and basis for stormwater credits.

Cost-effectiveness is a key reason stormwater utilities sometimes provide credits. The credits in these cities are often based on the estimated volume of retention/detention of the on-site BMP for a criteria storm. In other cases, credits against the stormwater utility fee are awarded if the BMP meets the standards for BMPs on development or for pollutant removal efficiency. The eligibility guidelines often rule out small or residential parcels, indicating that these cities believe the life-cycle costs of monitoring and maintaining numerous small-scale BMPs are prohibitive.

Credits might be a way to gain public acceptance of stormwater utility fees. By making stormwater utility fees a matter of choice, credits can reduce public opposition to utility fees. ILFs are similar in that they give landowners a choice between BMP installation or paying the ILF and may reduce political opposition to stringent on-site BMP requirements.

The literature review contributes to our understanding of cost-effectiveness. It is likely that on-site BMPs on small properties have high life-cycle costs because maintenance and monitoring costs are high. Cost-effective in-lieu fees will likely mandate in-lieu fees for these classes of properties. The review of stormwater credits shows that although credits are often structured for cost-effectiveness (by issuing credits based on the volume of runoff retained), a major purpose of credits could be to gain political acceptability. Stormwater credits are the closest policy instrument to ILFs, thus it will be interesting to observe if they are implemented similarly.

Survey Methodology

To fill in the gap in the stormwater literature on ILFs, we conducted a national survey of existing ILF policies. Twenty-nine cities completed the key portions of our written survey on a website or by e-mail, and follow-up phone interviews were conducted with twelve of these cities.*

It is important to understand that our survey is not a representative sample of cities that have ILFs. Rather, it is the largest sample we could obtain given available time and existing information. We first searched the Internet and stormwater literature for any mention of cities that implement ILFs. We then asked survey respondents to

---

* Twenty-eight cities completed all questions.

TABLE 9.1

Survey Reponses

Survey Status	Number
Completed survey from city with active policy	28
Incomplete survey	2
Policy terminated	2
Policy never used	1
Policy under consideration (not yet implemented)	5
Weren't aware they had policy	3
Never had policy	2
Have other type of in-lieu fee policy (stormwater quality, wetland)	2
No definitive response	57
Did not locate survey respondent	73
Total	175

list other cities and counties they knew to have ILFs, thus identifying additional municipalities. This sampling method underrepresents cities that do not have their municipal code or stormwater manual posted online, and thus likely underrepresents smaller, poorer, less tech-savvy municipalities.

In total, 175 municipalities were identified as likely to have an ILF in their municipal stormwater policy. An e-mail containing an introduction to the project, a link to a web-based survey, and a text copy of the survey was sent to the city or county employee judged most qualified to complete the survey. These surveys were sent to 102 cities where we were able to identify an employee responsible for stormwater management. See Table 9.1.

Survey Findings

Our survey included questions concerning the adoption of ILFs, the basis for fee calculation, and the use of fees. The responses to these questions allow us to form preliminary conclusions on whether ILFs are intended and structured for cost-effectiveness, and the prospects for adopting this promising policy more widely.

Why Do Cities Adopt In-Lieu Fees?

Geographic clustering appears to be a major factor influencing the adoption of stormwater ILFs in our sample of cities. Figure 9.1 shows our widest measure of cities with ILF policies; the map shows all 175 cities that have a mention of an ILF on their website or in other stormwater literature. The map reveals that likely ILF cities are clustered in several regions.

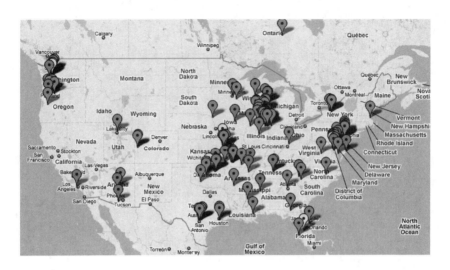

FIGURE 9.1
Each pin on this map represents one of the 175 cities that had, previously had, or was considering an in-lieu fee at the time of research (2008). A Google Maps version of this map can be viewed at: http://tinyurl.com/stormwaterILFs

The Chicago metro area contains 33 of the municipalities confirmed or suspected to have an ILF. Monroe County, New York (near Rochester) contains 26 municipalities with an ILF. Other smaller clusters are in the Washington–Baltimore corridor area, the Springfield, Missouri metro area, the Portland, Oregon metro area, the Seattle area, and the Kansas City area. It is striking that there is only one city in California, and only one municipality in New England outside Monroe County confirmed or suspected to have an ILF despite the large population of these areas.

This clustering of ILF cities is likely due to state or local policy promoting stormwater ILFs. Monroe County, New York is a cluster because every city adopted the county model code, which contained an ILF fee. DuPage County, Illinois had an ILF in their county model code. In addition, the Northeastern Illinois Planning Commission's 1990 Model Stormwater Drainage and Detention Ordinance contains an ILF. Maryland has long been a center for innovations in stormwater management.

The survey found the main reason for offering an ILF was the belief that in at least some circumstances regional stormwater management was more effective than on-site BMPs (see Figure 9.2). Another popular response was that on-site management was not appropriate in some circumstances. Varieties of circumstances were mentioned such as floodplain location, development in older areas with no setbacks, and small plots.

A less expected result was that 7 out of the 31 cities that answered the question cited the desire to raise on-site standards as a reason they

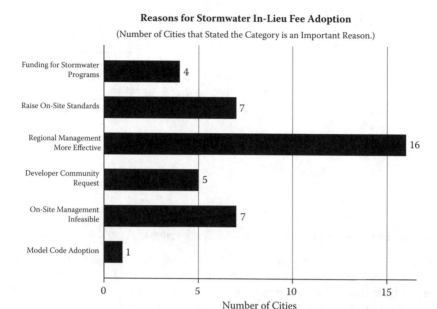

FIGURE 9.2
Reasons for stormwater in-lieu fee adoption.

adopted an ILF.* It appears that for a substantial minority of cities ILFs were adopted as a political safety valve to make adoption of on-site standards more palatable to the community. In addition, five cities added the ILF because of a request from the developer community. This can be interpreted as a public acceptance tactic. Finally, most cities (23 out of the 30 cities responding to this question) had more stringent stormwater requirements than their state. This suggests the cities with ILFs (and responding to the survey) are ones that are going a step beyond normal municipal stormwater requirements, and that fees provide necessary regulatory flexibility for more stringent on-site requirements.

These results suggest that ILFs are not just adopted as a cost-effectiveness tool; they play an important role in convincing the public to accept on-site BMP requirements. This is similar to the public acceptance role played by stormwater credits in convincing the public to accept stormwater utility fees. A final point is that the adoption of stormwater ILFs appears to be a policy that is gaining steam; 18 of 30 cities responding adopted ILFs in 2000 or later.

---

* Cities were allowed to select all categories that applied for question three, and then added some of their own reasons in additional comments, so there is more than one reason for adoption per city.

TABLE 9.2

In-Lieu Fees Vary Considerably Across Cities

	Impervious Area Basis (per Square Foot)	Storage Volume Shortfall Basis (per Cubic Foot)
Mean	$ 1.73	$1.96
Median	1.00	2.32
Min	.14	.23
Max	8.42	3.22
Number of Cities	9	6

Are In-Lieu Fees Structured for Cost-Effectiveness?

Our analysis of the adoption of ILFs suggests that cost-effectiveness is a major reason behind their adoption. However, are ILFs structured to maximize the cost-effectiveness of stormwater management? Our survey included questions on the structure of the ILF fees that can shed light on this question.

At least six cities base their fees on the volume of runoff, consistent with cost-effectiveness principles (see Table 9.2). The most popular fee calculation method, used by just under half of the municipalities surveyed, is to calculate the fee based on the square feet of added impervious or disturbed areas. Because impervious areas are a good proxy for estimating additional runoff from development, this method could be a second-best solution to basing the ILF on the volume managed by the required on-site BMP. Both fee calculation methods could be structured so the ILF is approximately equal to the marginal cost of regional facilities.

However, ILFs based on the cost of the BMP that would have been installed, a method used by several municipalities, are not consistent with cost-effectiveness. This method requires a site-specific BMP cost calculation where the developer pays the estimated cost avoided, or the developer pays an average cost for a typical BMP. The key shortcoming of both methods is that they are not tied to the cost of expanded regional infrastructure. If the cost of the BMP does not fully cover the cost of the incremental regional capacity needed to substitute for the lack of the on-site BMP, then allowing the ILF will decrease the overall cost-effectiveness of stormwater management.

A second difficulty with the site-specific BMP ILF calculation method is that any calculation of this type is likely to skip important cost components. Logically, the developer would only go through the time and expense of the ILF process if the ILF costs were lower than on-site placement. The opportunity cost of land and the cost in lost property value of installing BMPs are not directly observable, so these costs will likely not be fully captured by any ILF method based on avoided cost.

Our survey did not ask directly about whether the fee levels are based on average or marginal costs. It would take a detailed case study investigation to examine this question. However, the more expansive responses to the fee basis questions suggest that average and historical costs are the basis for many fees. Several responses stated the fee levels were based on historical costs or regional cost per volume detained, which is likely an average cost figure. Setting costs based on historical or average costs will usually result in ILFs that are too low and do not fully reimburse cities for the lost stormwater management capacity.

Developments are typically only eligible for ILF fees if they meet certain restrictions. Occasionally, these are minor restrictions, such as that ILF fees are only available if there are current downstream facilities. However, eligibility criteria are more commonly quite strict. Often, the ILF option is only available to small developments, or where construction of on-site BMPs is infeasible due to soil or drainage issues.

We expected to find that small and residential parcels were not eligible because of the high life-cycle costs of on-site BMPs for these land uses. However, the average city in our sample has eligibility requirements stricter than just prohibiting small and residential uses. We classified ILF eligibility requirements as low if: (a) applications are considered on a case-by-case basis, (b) the only requirement is that nearby properties are not adversely affected by the additional runoff from the proposed development, or (c) that downstream facilities must already be in place. By this measure, only 7 out of 29 cities have low barriers to ILF eligibility. In addition, several of those low requirement cities require variances from the city council or detailed study, so approval is by no means automatic. The overall strict limits on ILF eligibility—such as only allowing ILFs when it is proven that BMPs cannot be built—limit the cost-effectiveness potential of ILFs because many high-cost projects where ILF fees would be preferable are not eligible.

The surveyed cities largely allow partial detention plus a fee for the remainder (21 out of 29 cities). These are cases where developers install BMPs to manage a portion of the required volume or capacity, but there is a gap between the installed and required capacity. Allowing partial detention and a fee for the remainder is key for cost-effectiveness. Many properties likely have some areas where the costs of placing stormwater BMPs (with the opportune cost of land and lost opportunities for building) are fairly low, but costs quickly increase when the BMP infringes on land that is needed for parking or buildings. Allowing developers partially to pay their way out of detention combines the advantages of managing the runoff on-site while gaining revenue for regional programs.

Allowing partial detention is especially useful because the last bit of detention (whether measured by capacity or volume) is less valuable than initial capacity. The initial capacity serves to manage both small storms and the initial portion of large storms. However, the final portion of capacity is only used to manage a portion of larger storms. Where the concern is only peak flow from a criteria storm, this is a distinction without a difference: downstream facilities have to be designed to handle peak flow and therefore it is the peak flow handled by an on-site BMP that is important. However, in cases where water quality and cumulative impacts are a concern, the last increments of on-site capacity are less valuable than the initial increments. Where this is the case, ILFs should have a sliding scale where it is fairly cheap to pay a fee for the last increments of required management, but it becomes increasingly expensive to pay off the middle and initial increments. This would translate into block (tiered) rate fees where the initial increment of runoff volume difference between the required on-site BMP and the partial detention would be charged at one rate, and then the next increment of volume would be charged at a higher rate. There could be several tiers of rates depending on the needs of local stormwater management.

Discussion

Our analysis shows that the major reason for the adoption of ILFs is cost and effectiveness. Cities felt that in some circumstances, regional stormwater management will result in better results per dollar than on-site BMPs, and in those circumstances ILFs are a good policy.

However, most ILF policies are not set to maximize the cost-effectiveness of the runoff management system. Although many of the cities set volume-based fees that are consistent with cost-effectiveness, it appears that most cities are setting fees based on average and historical costs. Furthermore, few cities adjust fees to account for inflation rates. This practice likely results in an ILF that does not cover the full marginal cost of regional stormwater management.

The low ILFs may explain the very high eligibility requirements. If the fees do not cover the full marginal costs, agencies tasked with solving runoff quality and quantity problems will seek to limit the use of ILFs to the land uses and property development situations where they come closer to marginal costs. There may be a cultural element in explaining the very high eligibility requirements for ILF fees as well. Price-based policy instruments are often not intuitive for noneconomists and it is unlikely that the typical stormwater agency manager has much training in the nuances of these policies. Education and model ILF codes could contribute to a greater role for ILF and other price-based policies such as stormwater credits or subsidies.

High eligibility requirements, whatever their reason, are the greatest barrier to using ILFs to improve the cost-effectiveness of stormwater management. As currently structured in most cities, ILFs provide some flexibility for properties where on-site BMPs are infeasible, very costly, or difficult to maintain over the long run. Because they are not available to most properties, ILFs as currently structured are not likely to improve the overall cost-effectiveness of stormwater management systems to a significant extent.

It is easy to think that the way to remedy this situation is to raise fees until they cover the marginal cost of regional stormwater management and ease eligibility requirements significantly. However, this solution is too facile; the difficulty remains that after constructing regional stormwater detention systems based on planned and existing development in an area, it is very difficult to add more capacity. Physical constraints and local political and budgetary consideration could make expansion difficult even when it would be less expensive than on-site BMPs. Stormwater managers, under strict regulatory requirements, would then be understandably hesitant to give up the bird in hand of on-site capacity to speculate on regional capacity that has not yet been developed.*

The way out of this dilemma is to recognize there are other options in addition to on-site BMPs on new or redeveloped and centralized regional programs. Development of ILFs should be guided by the idea that they can fund a range of programs, including: runoff management in public-right-of-ways such as streets and sidewalks; on-site capacity in existing built-out areas, including residential properties; and expansion of runoff management capacity at new or redeveloped sites with relatively inexpensive opportunities for runoff management.

Runoff management in the public right-of-way has shown promise with the Seattle Sea Street projects. Projects in the Los Angeles area such as Oro Street and Elmer Street were built on the Sea Street concept and adapted to local conditions (Los Angeles San Gabriel River Watershed Council, 2010). Although not applicable in all areas, there are likely areas where ILFs could fund cost-effective street or sidewalk projects.

Thurston et al. (2008) analyzes a field experiment offering incentives for homeowners to install rain gardens and rain barrels in a Cincinnati suburb and finds that homeowners are willing to install these proven BMPs at a low

---

* Unfortunately, questions about complex motivations are not easily answered by surveys. This observation is based on discussions with Los Angeles-area stormwater managers.

cost. Cutter et al. (2008) analyzes an incentive-based system for retrofitting commercial and industrial properties in the Los Angeles area and finds there are inexpensive stormwater retrofit options that can be spurred by incentives. Both of these could be effective uses for ILF funds in some locations.

In addition, ILFs could be altered so that they provide subsidies for additional runoff management for new development and redevelopment projects in which runoff management beyond what is required would be cost-effective. Just as some development is likely to have very high costs of installing on-site BMPs, other development is likely to have quite low costs. If this development were in areas that need runoff management capacity, it would be efficient to offer the developments subsidies to install additional capacity. These subsidies could be priced in much the same way as ILFs: at approximately the marginal cost of regional management alternatives. Once cities have implemented correctly priced ILFs, it should be relatively easy for them to implement other incentive-based programs.

Developing these alternatives to the binary on-site versus regional management choice is key to taking advantage of the full potential of ILFs. If stormwater managers have a number of choices to meet regulatory requirements, easing eligibility requirements will seem less risky. Of course, it is also necessary to correctly price ILFs so they reflect the marginal cost of alternative stormwater management strategies.

Conclusion

In-lieu fees are a promising policy to move stormwater management to a more efficient system where runoff management is achieved at the lowest overall cost. ILFs allow developers and property owners to pay for regional treatment funding instead of managing runoff on-site at sites where on-site BMP installation is not cost-effective. At present, ILFs do not appear to be widely implemented.

As runoff and water quality regulation become increasingly stringent, it is likely that more cities will turn to tools such as ILFs to manage the overall costs of meeting federal and state runoff requirements. Our survey results indicate that ILFs are useful in gaining public acceptance of higher on-site standards. The surveyed cities have made a promising start at implementing ILFs that move regulation toward greater cost effectiveness.

Unfortunately, the strict eligibility requirements of many ILF policies limit these policies' overall effectiveness. In most of the cities we surveyed, a significant proportion of new development would not be eligible for ILFs. In some cities we surveyed, it appears that only narrow classes of new development would be eligible. This greatly limits the ability of

municipalities to use ILFs to generate revenue for more cost-effective programs. The strict eligibility requirements may be the result of too-low fee
levels, in turn due to the use of average rather than marginal costs to set
these fees. Fees should be set to reflect marginal costs of regional treatment and should be updated annually to take into account inflation in
construction costs.

Our final suggestion is that plausible alternatives to large regional
treatment such as incentive-based methods and runoff management
in public right-of-ways should be more fully developed. These incremental programs can be phased in as ILF revenues become available
so that stormwater managers do not have to be concerned about being
out of compliance with regulatory requirements while waiting out the
long lead times for construction of large regional runoff management
infrastructure.

References

Center for Watershed Protection, Stormwater Manager's Resource Center. (2008).
Model Post-Construction Stormwater Runoff Control Ordinance. Retrieved
February 10, 2009 from http://www.stormwatercenter.net/Model%20
Ordinances/Post%20Construction%20Stormwater%20Management/
Final%20Model%20Stormwater%20Control.htm

Commission, N.I.P. (1990). Model Stormwater Drainage and Detention Ordinance:
A Guide for Local Officials. Chicago.

Cutter, W. B., K. A. Baerenklau, A. DeWoody, R.S. Sharma, J.G. Lee. (2008). "Costs
and benefits of capturing urban runoff with competitive bidding for decentralized best management practices." Water Resources Research 44, W09410,
doi:10.1029/2007WR006343.

Debo, T.N., and A.J. Reese. (2003). *Municipal Stormwater Management.* Boca Raton,
FL: Lewis.

Doll, A., P. Scodari, and G. Lindsey. (1998). Credits as Economic Incentives for
On-Site Stormwater Management: Issues and Examples. *EPA National
Conference on Retrofit Opportunities for Water Resource Protection in Urban
Environments, Chicago,* February 9–12, 1998.

Dye, R.F., and D.P. McMillen. (2007). Teardowns and land values in the Chicago
metropolitan area. *J. Urban Econ.,* 61(1): 45–63.

Ferguson, B.K. (1998). *Introduction to Stormwater: Concept, Purpose, Design.* New
York: Wiley.

Jenkins, J.R., and E.E. Herricks. (1995). *Stormwater Runoff and Receiving Systems:
Impact, Monitoring, and Assessment.* Boca Raton, FL: CRC Lewis.

Los Angeles and San Gabriel Rivers Watershed Council. (2010). Water Augmentation Study: Research, Strategy, and Implementation Report. Los Angeles, California. Retrieved October 9,2010 from: http://lasgrwc2.org/dataandreference/Document.aspx?search=48

Maryland Department of the Environment. (2008). Report on Stormwater Management Act of 2007. Baltimore. Environment.

Moffa, P.E. (1996). *The Control and Treatment of Industrial and Municipal Stormwater.* New York: Van Nostrand Reinhold.

Nelson, A.C. (1995). *System Development Charges for Water, Wastewater, and Stormwater Facilities.* Boca Raton, FL: CRC Press.

Thurston, H.W. (2006). Opportunity costs of residential best management practices for stormwater runoff control. *J. Water Resources Plan. Manage.-ASCE,*132(2): 89–96.

Thurston, H.W., H.C. Goddard, D. Szlag, and B. Lemberg. (2003). Controlling storm-water runoff with tradable allowances for impervious surfaces. *J. Water Resources Plan. Manage.-ASCE,* 129(5): 409–418.

Thurston, H.W., A.H. Roy, M.A. Morrison, M.A. Taylor, and H. Cabezas. (2008). Using Economic Incentives to Manage Stormwater Runoff in the Shepherd Creek Watershed, Part I. Cincinnati, OH, National Risk Management Research Laboratory, U.S. Environmental Protection Agency.

Virginia Department of Conservation & Recreation. (2001). Virginia Model Ordinance for the Control of Post Construction Stormwater Runoff. Retrieved April 20, 2009 from http://www.dcr.virginia.gov/soil_and_water/documents/swmmodord.pdf

10

Cap-and-Trade for Stormwater Management

Haynes C. Goddard

CONTENTS

Introduction

This chapter is focused on the description and analysis of how a constructed pollution trading program, or cap-and-trade in popular language, might be applied to a stormwater management program to achieve runoff reduction goals at least cost and at the same time be reflective of property owners' preferences and opportunities.

Chapter 1 in this volume hinted at the problems caused by combined sewer overflows (CSOs) and storm sewer overflows (SSOs) that motivate much of the need to have more effective methods for dealing with runoff problems. Because in any one watershed there may be several thousand individual sources, stormwater managers have found, as we have seen in previous chapters, the traditional centralized gray infrastructural approach to managing wet weather flows can be expensive (such as stormwater tunnels). This has led to increased interest in and use of "green infrastructure," termed "best management practices" or BMPs, which are described elsewhere in this volume. There can be potentially thousands of BMPs on public and private properties in a stormwater management district; coordinating the placement and operation of these is daunting from a typical command-and-control perspective. What is needed is an efficient and fairly decentralized mechanism to carry out this coordination.

Pollution markets, when properly organized, are a great potential coordinating mechanism, and have been successfully employed in the United States to assign control responsibilities and investments for controlling SO_2 emissions. Therefore, there is a potential here to be explored in the stormwater case. A cap-and-trade program could identify the size and location of individual and regional level investments, and provide a means of financing them.

Background for Cap-and-Trade Programs

For any good or service to be exchanged in a market, the property rights to that good or service must be established, represented in a title or contract, be measurable, be exchangeable in a transaction, and be legally enforceable against theft or infringement. Without these conditions, markets cannot exist. In pollution trading, the property right traded is a discharge permit or "allowance" that measures one standard unit of discharge or emission.

Pollution trading is not about "free markets," as much experience with unfettered markets makes clear that unregulated markets are socially undesirable. Well-functioning markets, however, are socially beneficial and thus desirable, so the role of the economist is to describe the structure and operation of the relevant market and the required conditions that will produce improved social and economic welfare. With pollution trading, this means achieving pollution reduction targets at the least cost by creating a decentralized and voluntary exchange mechanism that allocates control responsibility to those who have lower cost options, and are willing to increase their levels of runoff control. This result is carried out by an early distribution of allowances that are less than all dischargers' current levels of discharge, and then offering them the opportunity either to control their own discharges on site, or to buy another's allowances that lead the seller to increase its control of its discharges beyond its initial allocation or endowment (Tietenberg, 2006).

The basic questions to be answered in constructing such a market are two:

1. What properties should the prices produced by such a market have that will lead to an efficient (least-cost) sharing of control responsibilities across all dischargers?
2. What is the practical design of such a market that will produce these efficiency prices?

The first question deals with explaining what the properties are of a well-functioning market in the stormwater case, and the second question with the features of an actual market. Finding answers to these questions is the focus of this chapter, so in the first part of this chapter we outline the framework in which these questions can be answered, followed by a discussion of several practical issues that must be attended to before such a constructed market can be put into operation.

Basic Conceptual Framework: Gray and Green Infrastructure

The traditional approaches to stormwater management have been to construct various types of underground structures, first to convey storm water and later human waste as well, and still later in the twentieth century to treat it (http://www.sewerhistory.org). This has always been the focus of the field we now call "sanitary engineering," with a focus on conveyance and treatment.

Economists call such infrastructure manufactured capital. With the rise of environmental concerns in the last 50 years and more recently environmental sustainability, ecologists and biologists have increasingly called attention to "nature's services" of pollutant degradation and assimilation, which is denoted as "natural capital" by ecological economists. To use the current terminology in the stormwater management field, manufactured capital is gray infrastructure, and natural capital is green infrastructure, mainly infiltration and evaporation.

There are several dimensions to natural capital or nature's services for stormwater management detailed elsewhere in this volume, and so here we mention two of the principal ones: evaporation/infiltration and flood storage. The following discussion assumes that both gray and green infrastructure will be used together in a least-cost stormwater management program. A market for trading runoff allowances is designed to provide answers to exactly what the least-cost combination is in practice.

The stormwater manager's problem is to manage stormwater flows at least cost, so this is a cost-effectiveness problem. With runoff, the current focus of the field is to reduce flooding and SSOs, and to eliminate CSOs. We capture the cost-effectiveness problem confronted by stormwater managers in Figure 10.1.

The diagram is based on the microeconomic concept of a production function, a mathematical relationship between inputs and output. To keep this in a two-dimensional plane, we denote runoff reduction as the output, produced by using a combination of two inputs: natural and manufactured capital, or gray and green infrastructure. There is both complementarity and substitutability between gray and green infrastructure, meaning that although they have to be used together, the input ratios are variable.

Figure 10.1 represents the cost-minimization problem as one of combining the inputs of (1) dispersed natural capital and (2) centralized infrastructure that meets stream flow and CSO/SSO constraints at least cost, at the watershed level. The stream flow constraint has to be transformed into a runoff goal or target through clear hydrological modeling of the watershed to be managed. The targeted runoff is represented as an iso-runoff function that can be met with various combinations of the two inputs; two possible process rays or expansion paths are shown. The iso-cost loci shown represent the relative prices (costs) of dispersed and centralized approaches to runoff management. This model shows the usual situation

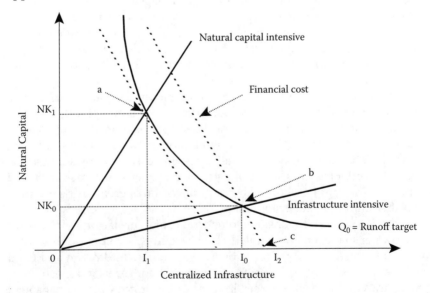

FIGURE 10.1
Least-cost combination of infrastructure and natural capital.

in which natural capital and manufactured infrastructure are simultaneously complements and partial substitutes.

In this chapter we hypothesize the traditional and typical, but now changing, emphasis on off-site approaches to managing runoff that cause stormwater management costs to be at, or to the right of, point b (i.e., higher cost) with infrastructure of I_0. Or the costs could be to the right of I_2, such as on a ray coincident with the horizontal axis that makes no use of dispersed retention, representing a lower level of dispersed runoff management and relatively greater investment in centralized infrastructure.

Point b is an intersection of the budget locus labeled "Financial cost" and the iso-runoff curve. This is not a least-cost solution as the first-order condition for a least-cost combination of the two inputs is achieved at a tangency on a lower iso-cost function at point a with the input combinations of (NK_1, I_1), which is relatively more natural capital-intensive than at point b. The stormwater manager's problem then is to find the combination of runoff abatement technologies (reduction in demand for sewerage from the installation of BMPs) plus centralized treatment and associated sewer infrastructure that minimizes the overall cost of reaching a set of environmental or ecological targets, such as flood prevention and preservation of stream habitat.

In this problem, the targets are initially defined as a runoff regime, measured in cubic feet per second, that can (1) enter streams without destroying stream habitat and creating flooding, and (2) enter the sewers and treatment systems without creating SSOs and CSOs. This choice of constraint serves to keep the initial analysis simple, but in fact the problem of urban runoff and CSOs is principally one of an excessive volume of water entering the sewer system, surface streams, and treatment facilities in short time intervals, a problem of peak flows that needs to be modeled stochastically.

Preservation of in-stream and riparian habitat requires in some cases that a minimal flow regime (base flow) be maintained. To focus on the essentials of the argument, the following exposition is in terms of runoff reduction only, and not water quality standards per se. However, runoff reduction will reduce pollutant levels in streams, and therefore contribute to achieving water quality objectives, perhaps substantially so in some watersheds.

With these assumptions, the constrained objective function is simple: minimize the total *social* cost (private plus public) of investment in:

1. Abatement or runoff reduction technologies (BMPs) and associated land area, plus

    2. Public infrastructure to convey and treat commingled waters, subject to a constraint on stream flow. Hydrological modeling determines the parameterization of the constraint(s).

Water Balance and Ecological Constraint

As noted, the basic underlying model that supports cap-and-trade programs is cost minimization subject to an emissions or ecological constraint. The constraint in this case is a limit on runoff, which we derive from a hydrological water balance. A standard hydrological description (Dunne and Leopold, 1978, p. 237) of *predevelopment*, deterministic water balance for a small drainage basin in a watershed (no ponding, lakes, sewers, nor flood storage explicit at this point) is given, with altered notation, by

$$R = I + AET + RO + \Delta SM + \Delta GWS + GWR \tag{10.1}$$

where all variables expressed in equivalent volumetric units per time interval and

R = rainfall;
I = interception (e.g., leaf canopies);
AET = actual evapotranspiration;
RO = surface runoff (denoted as overland flow or OF in Dunne and Leopold (1978));
ΔSM = change in soil moisture;
ΔGWS = change in groundwater storage;
GWR = groundwater or subsurface runoff.

Solving this for the surface flow, RO, gives:

$$RO = R - I - AET - \Delta SM - \Delta GWS - GWR \tag{10.2}$$

or simplifying,

$$RO = R - Detention / retention \tag{10.3}$$

where *Detention / retention* $= I + AER + \Delta SM + \Delta GWS + GWR$. *Detention/ retention* of surface runoff is denoted as DR^S in this development, giving Equation (10.3) as $RO = R - DR^S$. We note that in this exposition, groundwater runoff and the consequent base flow in receiving streams is not a management target, although in some areas that depend on groundwater for drinking water, it could be added to the problem. The focus here is on problems of excess stream flow arising from short-term storm events.

Formulating a constraint from the water balance, we have

$$R - DR^S(BMP) \le \overline{RO} \tag{10.4}$$

where the green infrastructure to promote *DR* is expressed as a function of the required *BMPs* to satisfy the constraint.

Expanding this with the gray and green infrastructure investments that are the stormwater manager's decision variables, Equation (10.5) expresses the constraint in terms of control zones (*z*), lots (*l*), a zonal parameter, b_z, reflecting zonal hydrological characteristics (slope, vegetative cover, soil type, etc.), and the associated gray infrastructure, denoted here as centralized investment (*CI*), principally sewers and treatment plants.

$$R - \sum_{z=1}^{Z} \left[b_z \sum_{l=1}^{L} \left(DR_{zl}^S \left(BMP_{zl}, CI_{zl} \right) \right) \le \overline{RO}_z \right] \tag{10.5}$$

Cost-Effectiveness Formulation

This allows us to give a mathematical representation of this optimization problem at the property level:

$$\underset{DR_{\forall z, \forall l}}{Min}\ C = \sum_{z=1}^{Z} \sum_{l=1}^{L} C_{zl} \left[DR_{zl}^S \left(BMP_{zl}, CI_z \right) \right] + \sum_{z=1}^{Z} \lambda_z \left[R - b_z \sum_{l=1}^{L} \left(DR_{zl}^S \left(BMP_{zl}, CI_z \right) \right) \le \overline{RO}_z \right] \tag{10.6}$$

where
  C = cost of runoff control as detention/retention;
  DR^S = detention/retention or surface runoff control as a function of investments;
  BMP = detention/retention investments on private properties;
  CI = centralized infrastructure such as sewers, flood storage and wetlands;

z = hydrologic zone (roughly corresponds to stream reach or catchment, whichever is more binding);

l = lot or property in a zone;

b_z = zonal parameter reflecting zonal hydrologic characteristics: slope, vegetative cover, soil type, and so on;

$\overline{RO_z}$ = runoff target by zone.

A management zone is defined hydrologically, but also in terms of the location of the constraint in subwatersheds that have SSOs and CSOs, and is elaborated more fully below. Given the growth in connected impervious surface, the constraint is likely to be globally binding and so we presume interior solutions (no Kuhn–Tucker conditions). We find the typical first-order condition *within one zone*, defined as all flows *upstream* of an ecological constraint, for properties and centralized investments is:

$$\lambda_z = \frac{MC_{BMP_{zl}}}{\left(\dfrac{\partial DR_{zl}^S}{\partial BMP_{zl}} \right)} = \frac{MC_{CI_z}}{\left(\dfrac{\partial DR_{zl}^S}{\partial CI_z} \right)} \tag{10.7}$$

where the numerators are marginal costs and the denominators are the incremental runoff controlled (marginal products) by the alternative control mechanisms.

Equation (10.7) is a standard result for cost-effectiveness: allocate the investments in gray and green infrastructure so the incremental or marginal costs of runoff reductions are equalized across investments in the downstream and upstream zones in a subwatershed. Second-order or convexity conditions are assumed satisfied. This condition would apply to all management zones. The intuition behind this result is straightforward: one simply compares the cost of an individual gray or green investment to handle the next increment in runoff, and apply whichever is less costly. The point of cap-and-trade is to provide a decentralized mechanism that will generate those costs in explicit prices at explicit sites and by economic sector.

This fundamental result tells us the prices generated in a real-world cap-and-trade market should be equal to the incremental costs across all investment alternatives on both public and private parcels. The task facing the stormwater manager is to design a practical trading system that will produce that result. Next, we state without derivation more realistic properties for this result, and then turn to the issue of trading ratios and the structure of a real-world market for stormwater trading.

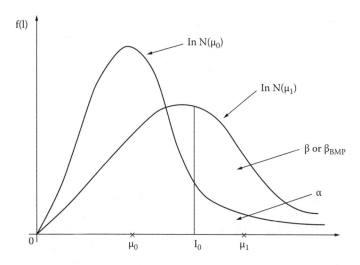

FIGURE 10.2
Stream flow density before and after control.

Stochastic Issues

The first obvious extension is that runoff is stochastic, following rainfall probability distributions. This means that controls focused only on average or mean runoff flows will not consider the runoff and stream flow variances, which in practice are flooding and CSO/SSOs. Avoidance of these excess runoff flows is the focus of much stormwater management.

Thus, the runoff constraint needs to be reformulated and expressed as a restriction on the likelihood or probability of exceeding some suitable measure of stormwater control performance, such as the frequency of bankfull stream flows and CSO/SSOs. We capture the basic ideas in Figure 10.2, represented as a frequency diagram in stream flow space, which requires hydrological modeling to translate Equation (10.7) that is expressed in runoff space.

The cost-effectiveness problem for the stormwater authority is two-part: choose the combinations of investments in *BMP* and *CI* (gray and green infrastructure) so as (1) to shift the mean μ of this distribution toward lower mean flows, and (2) to reduce the systematic part of the variance so I_0 is not exceeded more than α percent of the time. This is illustrated in Figure 10.2. Two density functions are shown: the one labeled $\ln N(\mu_1)$ is the precontrol distribution for stream flow index outcomes. If the stream flow norm is I_0 and the uncontrolled distribution of outcomes has a mean of μ_1, for example, then the norm I_0 is violated $(\alpha + \beta)$ percent of the time (area under the density function to the right of I_0) rather than the target frequency of exceedances of α percent (area under the curve labeled $\ln N(\mu_0)$ to the right of I_0). To meet the probabilistic standard, the control authorities choose and implement reductions in runoff through dispersed abatement and investments in centralized infrastructure that will shift the density function to that labeled $\ln N(\mu_0)$ so I_0 is not exceeded more than α percent of the time; β_{BMP} represents the reduction in the probability of exceedence through BMP investments.

This important modification of the constraint leads to a change in the cost-effectiveness condition presented above in Equation (10.7). We do not present a derivation here, but show what the first-order cost-effectiveness condition would look like with stochastic issues explicit. We shift the focus to include balancing investments across management sectors: household, commercial, and industrial sources and those investments the stormwater utility might make regionally, such as stormwater ponds, wetlands, and restoration of flood plains. Gray and green infrastructure choices are of course still involved here, but not explicitly to reduce the notational complexity.

The cost-effective balance of investments across sectors that considers the stochastic nature of surface runoff is expressed in Equation (10.8) with simplified notation:

$$\frac{\dfrac{\partial C^{HH}}{\partial DR^{HH}}}{\beta_{HH}} = \frac{\dfrac{\partial C^{COM-I}}{\partial DR^{COM-I}}}{\beta_{COM-I}} = \frac{\dfrac{\partial C^{PUB}}{\partial DR^{PUB}}}{\beta_{PUB}} \tag{10.8}$$

where the numerators are the areas labeled as β in Figure 10.2, the reduction in the probability of exceeding a stream standard for a design storm of given intensity and duration, and antecedent soil moisture conditions.

Again, the intuition is straightforward: given a runoff reduction target in a designated watershed, distribute the investments in gray and green infrastructure so the incremental or marginal cost per unit of reduced frequency of exceeding the target is balanced across the sectors and all parcels. Restated in traditional marginal cost (MC) notation this is given by

$$\frac{MC_{HH}}{\beta_{HH}} = \frac{MC_{COM-I}}{\beta_{COM-I}} = \frac{MC_{PUB}}{\beta_{PUB}} \tag{10.9}$$

Before we turn to issues of the design of a functioning market, we treat the issue of trading ratios that has been so prominent in the water quality trading literature.

Transfer Coefficients, Covariances, and Trading Ratios for Stormwater Flows

The closest related existing literature to the stormwater problem concerns point–nonpoint water quality trading. In that literature, transfer coefficients (delivery ratios) and trading ratios play an important, if problematic, role. Transfer ratios are designed to deal with the problem of varying concentrations of pollutants as they move through the environment, in principle making it possible to trade pollution allowances even though the impacts on receptors vary in time and space.

Transfer ratios are problematic because of the complexity they introduce into pollution trading markets (Hung and Shaw, 2005; Shortle and Horan, 2008; Sado, Boisvert, and Poe, 2010). However, as we show, transfer coefficients for stormwater trading alone (volume and flow reduction) can safely be ignored, thus reducing the complexity of an operational trading program, an important practical advantage.

Transfer coefficients or delivery ratios in standard deterministic modeling of the flow of an air or water pollutant through the environment account for pollutant accumulation and degradation as the pollutant is conveyed from location to location. These coefficients are fractions between zero and one that reflect the degradation and deposition of a pollutant as it is transported through the environment, so *ceteris paribus*, downstream and downwind damages from a given source diminish with distance. To understand the issues in the stormwater case, we need to expand the runoff relationship expressed in Equation (10.3) in the following manner.

Because the stormwater management problem is to manage stream flows for an entire watershed, we need to aggregate the flows. As groundwater runoff becomes base flow in streams, and we are interested in the volume of stream flow it moves through connected watersheds, this aggregation allow us to account for all water in the streams. Of course, during a wet season, groundwater flows will be high, raising the base flow to the streams and reducing the capacity of streams to receive new surface flows, which needs to be factored into the surface runoff constraints discussed earlier. We rewrite Equation (10.2) defining stream flow as the sum of surface and subsurface or groundwater runoff, or SF^{S+G}:

$$SF^{S+G} = RO + GWR$$

$$= R - I - AET - \Delta SM - \Delta GWS \qquad (10.10)$$

$$= R - DR$$

Moving from upstream to downstream, from zone 1 to zone Z, the exiting stream flow from the first control zone is the surface and groundwater runoff from that zone and is

$$SF_1^{S+G} = RO_1 + GWR_1 = R_1 - RD_1 \qquad (10.11)$$

For zone two,

$$SF_2^{S+G} = SF_1^{S+G} + R_2 - RD_2$$

and for the downmost zone

$$SF_Z^{S+G} = SF_{Z-1}^{S+G} + R_Z - RD_Z$$

or in general,

$$SF_Z^{S+G} = SF_{Z-1}^{S+G} + R_z - DR_z$$

$$= \sum_{z=1}^{Z} (R_z - DR_z)$$

$$= \sum_{z=1}^{Z} RO_z \qquad (10.12)$$

These relationships are needed to understand transfer coefficients in the stormwater case. Furthermore, these flows are of course correlated, an issue discussed below.

To analyze transfer coefficients or delivery ratios in the water balance model, we rewrite the expression between zones and make the coefficient or ratio explicit. This gives $SF_2^{S+G} = a_{12} SF_2^{S+G} + (R_2 - RD_2)$, where a_{12} is the transfer coefficient, indicating how much of zone one's runoff survives to reach zone two. We argue that this $a_{12} = 1$ in the stormwater case for the following reasons.

It is the *marginal* damage of the emissions from a single source that decreases as a function of distance. Of course if the total number of receptors along the pollution path increases, then the total damages for a given pollutant discharge can grow, although at a decreasing rate, reaching a maximum when the pollutant is either dissipated or consumed by stream biota. Note that even as the sum of the damages increases as a function of distance traveled, for pollutant trading, the trading ratio is established based on the marginal damage remaining in the zone from which a discharger wishes to increase discharges and who wants to buy a permit from an upstream discharger and permit seller. So in trading pollution rights, if the transfer coefficient is 0.5, the downstream buyer must buy two permits from the upstream seller, causing the upstream seller to reduce emissions by two units.

For stormwater, however, inasmuch as downstream locations always have had their own discharges independent of human actions, the stream channel has naturally increased in capacity to accept more stormwater. In contrast to air emissions or water pollutants, this capacity is automatically exhausted by the always existing runoff in the downstream locations, and so there is no added capacity naturally created as a function of stream length to accommodate increased runoff from upstream or downstream sources. In the case where development and channel modification have reduced stream capacity, this conclusion is only stronger. That is, reduced stream capacity would cause the marginal damages to increase at all points downstream of a discharge.

This leads to the conclusion that, because the focus here is on storm events, which are short term by definition, and because infiltration, evaporation, and evapotranspiration require time to operate, the transfer coefficient for storm-generated runoff should be treated as essentially equal to one (we do not model pollutant flows from CSOs here). This observation has major implications for stormwater runoff market design in that, without important differences in the *variances* of flows across trading areas or zones, it implies the trading ratio between upstream and downstream permits will be equal to one, important for keeping down the complexity of a trading mechanism. Finally, we note that should control of stormwater runoff be designated as a mechanism to manage water quality in a stream, trading ratios may become relevant.

Correlated Stream Flows: Covariances

As Equation (10.12) makes clear, the hydrological nature of catchments and watersheds correlates downstream stormwater flows with upstream flows, and it is the high flow covariances that result in stream damage and flooding. Thus, for uncontrolled stormwater there is no *a priori* reason to presume that variances will be constant across properties and subsets of properties in a given zone, nor that interzonal covariances will be zero. Furthermore, extending connected impervious surfaces resulting from development raises stream flow variability or flashiness and therefore covariances across zones.

Thus, positive stream flow covariances would seem to present a problem for stormwater trading by introducing the issue of nonunitary transfer coefficients and trading ratios via correlated flows. Although we have argued the transfer coefficients for stormwater are one, the fact that downstream flooding depends on the covariation of runoff from downstream and upstream zones would seem to undermine that argument. If so, the cost-effectiveness conditions derived above would need to be restated by including stochastic dependence across zones with joint probability distributions.

However, the objective of stormwater management is primarily to reduce two risks: (1) SSOs and flooding, and (2) CSOs. This is achieved by placing binding constraints on permitted runoff in all zones that contribute to the problems. These constraints are designed to eliminate covarying flows, causing the covariances that exceed stream capacity to be reduced to the probabilistic exceedances discussed above. That is, the problem of nonzero covariances between zones is controlled directly through establishing subwatershed stream flow constraints and the derivative property level runoff maxima. As a result, the trading ratio can remain at one.

With this exposition of the major conceptual issues, we turn now to discussing how a stormwater trading market would be organized.

Market Design and Operation

In this section we outline the major features of a practical stormwater trading mechanism, followed by a discussion of major areas of needed research. There are at least four major areas; we discuss the first two here.

1. Design alternatives
2. Setting up the market
3. Market operation
4. Funding and finance

Design Alternatives

The USEPA report "Water Quality Trading Evaluation" at http://www.epa.gov/evaluate/pdf/wqt.pdf identifies the following types of market transactions of relevance to water quality trading (p. 21), on which we briefly comment.

1. *Bilateral Negotiations:* Under this structure, each transaction requires substantial interaction between the buyer and the seller to exchange information and negotiate the terms of trade. Buyers and sellers make agreements on their own, with a public authority participating to approve the trade and set an appropriate trading ratio.

2. *Clearinghouse:* In this market structure, the link between the buyer and the seller is replaced by an intermediary. The clearinghouse is authorized by the oversight agency to pay for pollutant reductions and sell credits to sources that need them.

3. *Exchange:* This market structure is characterized by its open information structure and fluid transactions between buyers and sellers. In an exchange, the price for credits is fully visible. Exchanges can develop only when a unit of pollutant control from one seller is viewed as equivalent to a unit from any other source by quantity, quality, time, and space.

4. *Sole Source Offsets:* A sole source offset takes place when an individual facility is allowed to meet a water quality standard at one point if pollutants are reduced elsewhere in the same watershed, either on site or by carrying out pollution reduction activities off site.

5. *Third-Party:* In this market structure, buyers and sellers use a broker to conduct trading; the broker may be a regulatory agency, a nongovernmental agency (NGO), or an independent body established for trading. The broker facilitates bilateral trades; unlike a clearinghouse, using a third-party does not eliminate contractual or regulatory links between sellers and buyers.

Comments on Market Alternatives

Bilateral negotiations are always conducted when the good to be exchanged is nonuniform. Examples are new and used automobiles, houses, and real estate. This market structure may have a role in identifying regional stormwater BMPs, such as wet and dry ponds, where the parties to the negotiation would be the stormwater utility and an ad hoc association of runoff generators, such as neighbors in a contiguous watershed.

Clearinghouses are commonly used in the financial industry, but are not familiar to consumers. However, this form, combined with some aspects of the other market structures, is likely to be the best arrangement for stormwater trading.

An exchange is a familiar trading structure; examples are stock exchanges and commodity exchanges. Basic to the operation of an exchange is that the good be uniform in quality and type. Although stormwater is uniform as a commodity (abstracting from pollutants that it carries), it is not uniform in space; where it is generated and the locations of its impacts matter, and this was the motivation for discussing trading ratios above. However, once ecological and flood constraints are in place, trading stormwater reductions among sources will have some similarities to an exchange, making the good in question, runoff reduction, sufficiently uniform to allow trading.

Sole source offsets are commonly used for air pollution controls, where a firm desiring to expand operations and emissions has to find an existing source with the same type of emissions, which it will offer to abate directly by funding emissions reduction investments. This frees up environmental capacity for the new firm's emissions. These might have a role such as a new runoff source offering to install stormwater ponds off site on another property in the controlled watershed.

Third-party arrangements are through brokers. Brokers are widely used in the commodities markets, and in the case of SO_2 and NO_x allowance trading, brokers are commonly used. (See http://www.epa.gov/airmarkets/trading/buying.html#broker.) Brokering trades may have a role in stormwater trading, particularly among large sources.

Of these five market arrangements, the clearinghouse seems the most appropriate for stormwater trading; it could include elements of an exchange, bilateral negotiations, and sole source offsets as well. We next describe in broad terms how a stormwater trading market can be set up and operated, and then return to this issue. Inasmuch as there are no established stormwater trading markets anywhere, there will likely be much trial and error in identifying the best trading arrangements.

Establishing the Market for Stormwater Trading

The rationale for markets is of course to facilitate the exchange of goods and services, but what is actually traded in markets is the property right to the goods and services. A store receipt should be thought of as a title to the good, and what the transaction does is transfer the property right or title from the seller to the buyer. Critical to any market exchange is that property rights have to be established first, and have to be enforced, and that is where a stormwater market starts: the assignment of a property right for runoff, the possessor of which has legal access to off-site stormwater management services.

Legal precedent is well established to collect fees both for flows to the sewer system, and for installation costs for those properties switching from septic systems to the sewer network. The sewer district is the seller of sewer services, and the gray infrastructure is funded and financed through taxes and/or sewerage fees, commonly calculated as an average cost over the sewer district and added to the water bill.

For the case of stormwater, past practice has been for the property owner to have open access to gray infrastructure at a zero price for its runoff, either through direct connections, or indirectly through curbside discharges, both of which have the effect of placing a zero value on gray infrastructure and making green infrastructure uncompetitive even when it would result in lower stormwater management costs. It is important to note that a tax or flat stormwater fee is not the same as a price, which is a per unit charge.

From an economic perspective then, that is the source of SSOs and CSOs: the gray infrastructure is underpriced, leading to excess demand for it, and reduced demand for green infrastructure. Increasingly, however, in response to federal mandates to reduce water pollution for stormwater, stormwater utilities are charging flat-rate fees to property owners based either on the average runoff in an area, or based on estimated runoff from the calculated impervious surface on a site. Some stormwater utilities offer credits for certified runoff abatement (see, e.g., http://www.ci.minneapolis.mn.us/stormwater/fee/StormwaterQuantityCredits.asp in Minneapolis, Minnesota), which is an important step in the direction of establishing a price for runoff management services.

Several necessary steps in creating a stormwater trading market are:

1. Determine where in the regional stream system flooding and ecological damage limits are to be placed based on historical flood and ecological damage records. This is the underlying scarcity that drives limits on permitted runoff at the property level.

2. Establish legislation that limits stormwater runoff from properties. This is the cap of "cap-and-trade." For the purpose of illustration, a restriction that limits runoff to that which was generated from the site in its predevelopment (urban and agricultural) condition would be one that would generate strong incentives to install green infrastructure on both existing and new development. These limits would be based on site characteristics such as slope, soil type, and assumed vegetative cover, and a design storm of a given frequency, intensity, and duration.

3. Create a GIS database that shows the potential for individual and regional (neighborhood) BMP placement and the possible routing of runoff if different from existing routing. The USEPA program SUSTAIN (http://www.epa.gov/nrmrl/wswrd/wq/models/sustain/) is one such program that could be adapted for this purpose.

4. Distribute to existing property owners, free of charge, allowances for that amount, indicating these are transferable in a market transaction. This is the predevelopment endowment.

5. Establish a market structure that allows the property owner options for off-site control when he chooses not to install BMPs on his own property to detain and retain that runoff. These (among others) can be (a) an option to purchase an upstream discharger's allowances who can and is willing to detain more than her allowances permit, or (b) an option to buy shares or credits in neighborhood/regional BMPs such as streetside bioswales and ponds. Tradable allowances would be denominated in standard units, such as cubic feet.

6. The stormwater utility (SWU) would establish and operate a clearinghouse that would both buy and sell allowances in the various controlled watersheds, and would coordinate and install regional BMPs such as wet and dry ponds, flood storage, and constructed wetlands, as well as match investments in gray infrastructure.

7. The SWU would develop a menu of approved BMPs and license landscape architects and other contractors to analyze individual site requirements and BMP performance, to price, to install, to verify, and to insure/warrantee them for a stipulated time period. Examples can be seen at online site for Portland, Oregon, Bureau on Environmental Services (http://www.portlandonline.com/bes/index.cfm?c=31870), especially the 2004 Stormwater Management Manual.

8. The SWU would either provide directly and/or license landscape architects contractors to monitor, inspect, and maintain BMPs, all enforcement activities, similar to the current practice of building code inspections and compliance.

9. The SWU would either collect a surcharge on allowance trades or establish a site-based stormwater utility fee to cover overhead expenses.

General Comments

Before a market can function and generate positive prices, there has to be scarcity. By establishing limits on total runoff, access to sewers and streams will be made scarce to those seeking to manage their runoff off site and they will be motivated to seek lower cost alternatives and to increase the installation of green infrastructure.

The purpose of the predevelopment runoff ceiling is to establish limited access to sewers and streams in a way that makes clear the level of "nature's services" before the site was developed. This recognizes that stream channels and capacities were naturally established based on these predevelopment flows. Current stream capacities may in fact be less than these due to channel modification and loss of flood storage areas, and may require even tighter restrictions.

Imposing new limits will always be controversial, but such limits could be argued to be nonarbitrary in the sense the limits are independent of political influence. A property owner facing such a limit has essentially three choices: (a) install and maintain BMPs on the site to detain/retain runoff quantities exceeding the predevelopment endowment limit, (b) purchase an upstream site owner's excess allowances, or (c) join a neighborhood or regional association that requests the SWU to install and maintain neighborhood or regional BMPs such as ponds to detain/retain runoff flows exceeding the neighborhood's predevelopment flows. The SWU will sell allowances to the affected site owners, and use the funds to build and maintain the BMPs.

Most metropolitan areas now have GIS maps at the county level. These likely will have to be refined to measure lot perimeters accurately, encode soil type, slope, and vegetative cover and other relevant variables to compute the predevelopment runoff levels.

Concerning the initial distribution of allowances, although there is a substantial literature suggesting that economic efficiency (net benefits) is higher when the allowances are auctioned off, and revenues are generated in the process, political realities suggest that a free grant or grandfathering will be more successful in obtaining public support or acquiescence. Auctions will lead many to raise issues of regressivity with respect to income. Although that is not correct, as purchased allowances are substitutes for BMPs, such a perception will lead to increased opposition to the trading concept. The calculation of the endowment would be based on the hydrologically modeled predevelopment runoff.

The SWU would maintain a searchable database of the allowance endowments in the relevant watershed and of the possibilities for neighborhood/regional BMPs. The SWU would function as a clearinghouse, offering to buy excess allowances and to sell allowances for each sub-watershed subject to control.

A clearinghouse seems to be the most appropriate market structure for stormwater trading, perhaps incorporating some features of the other forms described above. A clearinghouse would match up potential buyers and sellers, ensuring that both are in the same watershed, collect the payments from the buyer, pay them to the seller, and ensure the seller of an allowance complies by installing and maintaining the BMPs.

A Stormwater Trading Narrative

This narrative reviews the various steps that a single residential property owner would need to address to make a decision to choose on-site or off-site stormwater control. It illustrates what information is required and how to interface with the SWU's clearinghouse to decide to buy or sell runoff allowances and credits.

Residential Properties

1. The property owner receives notice from the SWU that under the new stormwater utility and runoff management program only his predevelopment runoff, as modeled hydrologically, is eligible for access to the sewer system free of per unit charge. He receives a notice of the current runoff as modeled, the number of cubic feet of runoff for a design storm that are allowed from his property, which becomes his allowance endowment. Instead of an auction to finance the gray infrastructure, there will continue to be a monthly flat rate stormwater fee.

2. The SWU explains that the property owner has essentially three choices for managing his runoff: (a) detain and retain it on his property by installing BMPs, (b) join a neighborhood or regional association to participate in constructing a neighborhood or regional BMP that will accept and manage his runoff in exchange for purchasing allowances issued by the SWU, or (c) through the clearinghouse find an upstream property owner who offers to sell her allowances at an agreed-upon price.

3. The property owner is provided with an assessment of the soil type on his property, and a list of approved BMP methods and required capacities to detain and retain the amount of runoff that exceeds his endowment allowance. He receives a list of the range of average costs for installation of the BMPs by a certified and licensed landscape architect or contractor. A list of certified

installers is provided as well. The SWU provides a worksheet to allow the property owner to calculate the costs of the BMPs, installed either by a contractor or the property owner.

4. The SWU provides the property owner with the location and estimated costs to construct a neighborhood-level BMP such as a pond for off-site management or infiltration swales at roadside, based on stormwater routings, or a regional BMP for a collection of neighborhoods. The neighborhood BMP would be financed by selling additional allowances to those property owners who would prefer off-site management.

5. The SWU operates an online clearinghouse that allows buyers and sellers of allowances to post bids. The only restriction is they must lie in the same watershed, and the sellers of allowances must be upstream of the buyer. Once a bargain is struck, the clearinghouse certifies the sale, receives the funds from the buyer, transfers them to the seller upon completion and certification of the BMPs, and transfers the allowances. The seller of an allowance has a suitable amount of time to install the BMPs according to specifications the SWU determines. The plans and the final installation must be approved by the SWU, a process equivalent to building codes. This process could be contracted out by the SWU to the existing building code offices. BMP installers must guarantee the viable function of the BMPs for a time period specified by the SWU. The purchased allowance would be time limited to the expected life of the BMP, but could be renegotiated at the time of renovation should the parties agree.

Commercial, Industrial, and Other Properties

1. The process in this case is very similar to the above, but given the extent of the impervious surfaces involved, in addition to regional BMPs and individual sales, the possibilities of sole source offsets, bilateral negotiations, and brokerage are possible as well.

2. Allowance banking would be permitted and administered by the SWU.

Streets and roads are significant sources of runoff as well, therefore street and road departments could be given a predevelopment allowance, and would have to either directly install BMPs such as curb-cut rain gardens or participate in the trading market as well.

Conclusions

This analysis suggests the use of a stormwater trading program may well be part of a cost-effective stormwater management program. An appropriate next step would be to conduct a pilot test in a suitable watershed in a community that has serious CSO and SSO problems and is looking for lower-cost green infrastructure alternatives.

References

Dunne, T., and Leopold, L. (1978). *Water in Environmental Planning.* New York: W.H. Freeman.

Hung, M.-F., and D. Shaw. (2005). A trading-ratio system for trading water pollution discharge permits, *J of Environ. Econ. and Manag.* 49: 83–102.

Minneapolis, Minnesota, Storm and Surface Water Management, Applying for Stormwater Quantity Credits. Accessed at http://www.ci.minneapolis.mn.us/stormwater/fee/StormwaterQuantityCredits.asp

Portland, Oregon, Bureau on Environmental Services. Accessed at http://www.portlandonline.com/bes/index.cfm?c=31870),

Sado, Y., R.N. Boisvert, and G.L. Poe (2010). Potential cost savings from discharge allowance trading: A case study and implications for water quality trading. *Water Resource Res.,* 46, W02501, doi:10.1029/2009WR007787

Shortle, J.S., and R.D. Horan. (2008). The economics of water quality trading. *Intl. Rev. Environ. Resource Econ.* 2(2):101–133.

Tietenberg, T. (2006). *Emissions Trading: Principles and Practice,* 2nd ed. Washington, DC: Resources for the Future, Inc.

U.S. Environmental Protection Agency (USEPA). (October, 2008). Water Quality Trading Evaluation. Retrieved from http://www.epa.gov/evaluate/pdf/wqt.pdf

U.S. Environmental Protection Agency (USEPA) Clean Air Markets, Buying Allowances. (See http://www.epa.gov/airmarkets/trading/buying.html#broker.

Index

Page numbers followed by *f* or *t* refer to figures or tables.